Gaming Media and Social Effects

The scope of this book series is inter-disciplinary and it covers the technical aspect of gaming (software and hardware) and its social effects (sociological and psychological). This book series serves as a quick platform for publishing top-quality books on emerging or hot topics in gaming and its social effects. The series is also targeted at different levels of exposition, ranging from introductory tutorial to advanced research topics, depending on the objectives of the authors.

More information about this series at http://www.springer.com/series/11864

Yiyu Cai · Qi Cao
Editors

When VR Serious Games Meet Special Needs Education

Research, Development and Their Applications

Springer

Editors
Yiyu Cai
School of Mechanical and Aerospace Engineering
Nanyang Technological University
Singapore, Singapore

Qi Cao
School of Computing Science
University of Glasgow, Singapore Campus
Singapore, Singapore

ISSN 2197-9685 ISSN 2197-9693 (electronic)
Gaming Media and Social Effects
ISBN 978-981-33-6944-3 ISBN 978-981-33-6942-9 (eBook)
https://doi.org/10.1007/978-981-33-6942-9

This Springer imprint is published by the registered company Springer Nature Singapore Pte Ltd.
The registered company address is: 152 Beach Road, #21-01/04 Gateway East, Singapore 189721, Singapore

Foreword

When I was approached to write this Foreword, I was impressed to learn how Prof. Yiyu Cai's body of work (represented in this book) took roots from the 2003 Severe Acute Respiratory Syndrome (SARS) episode in Singapore. This subsequently led his team to employ virtual reality (VR) to convey the hand, foot, and mouth disease to be exhibited at the Singapore Science Centre. Moving from his work on diseases, it was almost a natural progression that the team extended their work to consider the support of children with autism spectrum disorder (ASD). The rest is as they say is history. At each progressive step, Prof. Cai received competitive grants to apply engineering solutions through VR as well as augmented reality (AR) to not only science but special education to help students with disabilities better access and engage learning.

As I began to understand VR and AR a little more, it became apparent to me how beneficial simulated environments could be for creating learning opportunities for students with disabilities. Medical researchers, for example, identified four key advantages of the use of simulation to train prospective medical personnel: (a) The teaching agenda can be focused on the needs of the students rather than relying on the patient needs to dictate the learning; (b) the simulated environment is a safe place for students to take chances and try new procedures; (c) educators can develop specific measures of success using metrics that might not be possible in live situations; and (d) learners can receive immediate feedback in a way that encourages maximum learning [1]. This list of reasons is persuasive, and the advantages are also applicable in educational simulation [2]. This being the case, the possibilities then for special education will be even more compelling.

Employing VR and AR has in recent times been adopted for promoting various educational and independent living supports [3, 4]. VR has positive potential for skill development in diverse areas as personal safety skills through serious game-based trainings and simulators [5], facilitating self-learning methods [6], simulating realistic experiences in high-risk safety training [7] and development of hand skills for surgical training [8].

In the areas of disability, VR was used to train social skills for individuals with mental health issues such as schizophrenia. Individuals using VR technology were reported to demonstrate greater improvement compared to traditional methods of teaching skills acquisition. Findings pointed to VR as a complementary means to support traditional social skills training to improve motivation and learning [9]. For individuals with ASD, VR was employed to teach emotional control and calming strategies to develop emotional and social adjustment. The study reported VR-based skill training improves the positive emotional control, emotional expression, and social–emotional interaction skills of children with autism [10]. VR was also used to train students with ASD in preparing them for daily living skills such as training for shopping using a head-mounted VR display to simulate a supermarket situation. The study reported benefits when participants were brought to the supermarket [11].

Similarly, the potential of AR applications provides opportunities for varying degrees of immersion, interaction, and involvement for disability-specific social services, i.e., independent living and physical and learning programs [12]. Further, the nature of multiple environments afforded by AR applications allows the merging and real-time interaction of virtual objects with real ones [13]. These features are especially pertinent in the field of special education as the discipline requires specialized teaching strategies to facilitate the learning of a broad range of functional skills to enable individuals to perform self-care activities at home, school, and work. These functional skills span broadly to include vocational skills, social skills, and behavior management skills [14].

For children with special educational needs requiring independent life skills training, the use of AR environment provides structured and enriched learning environment for conducive learning [15]. AR has also been used in training wayfinding skills for students with intellectual disabilities [16], math and numeracy [17], behavior management [18], literacy, and recreational skills [19].

From these specific advantages, further areas of benefits emerging from the tasks and skills gained include the strengthening of self-determination and self-management and improved guidance in self-instruction in complex tasks resolution, as well as guidance and location in various environments [20]. The benefits of AR also afford individuals with disabilities the opportunities to pursue interests and hobbies [3].

Where the literature points to an increasing number of studies in the area of VR and AR to support learning for students with disabilities, the application of these technologies to individuals with disabilities is still an emerging field with few papers published [20].

In response to this emerging area, the publication of this volume is a collection of experiences of the potentialities, affordances, and challenges of use of virtual and augmented worlds to enable learning and motivational games; teaching and learning simulations to address the curriculum needs as reflected in special education in Singapore.

The chapters share the evidence and potential for VR in order to narrow several gaps in the field: (a) the potential of VR and AR in special education, (b) the use of different virtual and augmented realities approaches in special education environments, and (c) technical affordances of virtual and augmented environments.

This book will contribute to the growing area of VR and AR as technology continues to complement special education to improve teaching and learning to better support the preparation and training of individuals with disabilities as they transition to society. This could be the nascent conduit between special education and technology for the local landscape. It is now up to both disciplines to deeper collaborate to build upon the strengths from each field to enable access to greater teaching and learning for the betterment of the lives of individuals with disabilities.

October 2020

Meng Ee Wong, Ph.D.
Associate Professor
Psychology and Child & Human Development
National Institute of Education
Nanyang Technological University, Singapore

References

1. Kneebone, R.: Simulation in surgical training: educational issues and practical implications. Med. Educ. **37**, 267–277 (2003). https://doi.org/10.1046/j.1365-2923.2003.01440.x
2. Spencer, S., Drescher, T., Sears, J., Scruggs, A., Schreffler, J.: Comparing the efficacy of virtual simulation to traditional classroom role-play. J. Educ. Comput. Res. **57**(7), 1772–1785 (2019). https://doi.org/10.1177/0735633119855613
3. Alshafeey, G.A., Lakulu, M.M., Chyad, M., Abdullah, A., Salem, G.: Augmented reality for the disabled: review articles. J. ICT Educ. **6**(46–57) (2019).
4. Baragash, R.S., Al-Samarraie, H., Alzahrani, A.I., Alfarraj, O.: Augmented reality in special education: A meta-analysis of single-subject design studies. Eur. J. Spec. Needs Educ. 1–16 (2019). https://doi.org/10.1080/08856257.2019.1703548
5. Backlund, P., Engstrom, H., Hammar, C., Johannesson, M., Lebram, M.: Sidh—a game based firefighter training simulation. 11th International Information Visualization Conference, Zurich, Switzerland (2007). https://doi.org/10.1109/IV.2007.100
6. Chittaro, L., Ranon, R.: Serious games for training occupants of a building in personal fire safety skills. International Conference on Games and Virtual Worlds for Serious Applications, Coventry, England (2009). https://doi.org/10.1109/VS-GAMES.2009.8
7. Smith, S., Ericson, E.: Using immersive game-based virtual reality to teach firesafety skills to children. Virtual Reality **13**(2), 87–99 (2009). https://doi.org/10.1007/s10055-009-0113-6
8. Aggarwal, R., Ward, J., Balasundaram, I., Sains, P., Athanasiou, T., Darzi, A.: Proving the effectiveness of virtual reality simulation for training in laparoscopic surgery. Ann. Surg. **246**(5), 771–779 (2007). https://doi.org/10.1097/SLA.0b013e3180f61b09
9. Park, K., Ku, J., Choi, S., Jang, H., Park, J., Kim, S., Kim, J.: A virtual reality application in role-plays of social skills training for schizophrenia: A randomized, controlled trial. Psychiatry Res. **189**(2), 166–172 (2011). https://doi.org/10.1016/j.psychres.2011.04.003
10. Ip, H., Wong, S., Chan, D., Byrne, J., Li, C., Yuan, V., Lau, K., Wong, J.: Enhance emotional and social adaptation skills for children with autism spectrum disorder: A virtual reality enabled approach. Comput. Educ. **117**, 1–15 (2018). https://doi.org/10.1016/j.compedu.2017.09.010
11. Adjorlu, A., Hoeg, E.R., Mangano, L., Serafm, S.: Daily living skills training in virtual reality to help children with autism spectrum disorder in a real shopping scenario. In 2017 IEEE International Symposium on Mixed and Augmented Reality (ISMAR-Adjunct), pp. 294–302 (2017). https://doi.org/10.1109/ISMAR-Adjunct.2017.93

12. Wu, H., Lee, S., Chang, H., Liang, J.: Current status, opportunities and challenges of augmented reality in education. Comput. Educ. **62**, 41–49 (2013). https://doi.org/10.1016/j.compedu.2012.10.024
13. Chen, C., Lee, I., Lin, L.: Augmented reality-based self-facial modeling to promote the emotional expression and social skills of adolescents with autism spectrum disorders. Res. Dev. Disabil. **36**, 396–403 (2015). https://doi.org/10.1016/j.ridd.2014.10.015
14. Stahel, A.: Daily living skills. In: Volkmar, F.R. (ed.) Encyclopedia of Autism Spectrum Disorders, pp. 839–840. Springer, New York (2013)
15. Cakir, R., Korkmaz, Ö.: The effectiveness of augmented reality environments on individuals with special education needs. Educ. Inf. Technol. **24**(2), 1631–1659 (2019). https://doi.org/10.1007/s10639-018-9848-6
16. Smith, C., Cihak, D., Kim, B., McMahon, D., Wright, R.: Examining augmented reality to improve navigation skills in postsecondary students with intellectual disability. J. Spec. Educ. Technol. **32**(1), 3–11 (2017). https://doi.org/10.1177/0162643416681159
17. Kellems, R.O., Cacciatore, G., Osbourne, K.: Using an augmented reality-based strategy to teach mathematics to secondary students with disabilities. Career Dev. Transition Except. Individ. **42**(4), 253–258 (2019). https://doi.org/10.1177/2165143418822800
18. Cakiroklul, U., Gokoglu, S.: A design model for using Virtual Reality in behavioral skills training. J. Educ. Comput. Res. **57**(7), 1723–1744 (2019). https://doi.org/10.1177/0735633119854030
19. McMahon, D., Cihak, D., Wright, R., Bell, S.: Augmented reality for teaching science vocabulary to postsecondary education students with intellectual disabilities and autism. J. Res. Technol. Educ. **48**(1), 38–56 (2016). https://doi.org/10.1080/15391523.2015.1103149
20. Gomez-Puerta, M., Chiner, E., Melero-Perez, P., Lorenzo, G.: Research review on augmented reality as an educational resource for people with intellectual disabilities. Int. J. Dev. Educ. Psychol. **3**(1), 473–485 (2019). https://doi.org/10.17060/ijodaep.2019.n1.v3.1523

Contents

Introduction

Yiyu Cai and Jieqiong Chen

Abstract This chapter introduces the motivation and background of developing VR and serious games technology for special needs education. This book is a collection of projects developed in Nanyang Technological University in collaboration with special needs schools in Singapore.

Keywords Education technology · Learning through playing

In 2003, Singapore was experiencing big difficulties due to the SARS pandemic. It happened that time the first author was leading a team in Nanyang Technological University (NTU) doing research on virtual reality (VR) and visualisation. The curiosity about the SARS virus had driven the team to design a VR roller-coaster game for the purpose to interactively visualise the internal structures of some SARS viral proteins [1]. This game was invited by Singapore Art Museum for an immersive and interactive exhibition at the Gallery #10 for a period from September 2003 to October 2004. Around 2005, two serious games for Hand Foot Mouth Disease and Influenza Virus were developed by the same team for a permanent exhibition in Science Centre Singapore.

In 2010, the first author and his team were awarded a seed grant from the Institute for Media Innovation at NTU to investigate the use of VR technology for children with autism spectrum disorder (ASD). Subsequently, the team received a funding support from Temasek Trust Funded Singapore Millennium Foundation to design a serious game called Virtual Pink Dolphins with an aim to assist children with ASD in their learning and communication [2]. Asia Women's Welfare Association (AWWA) School is a special needs school in Singapore among the first with the Virtual Pink Dolphins serious game used in classroom learning [3].

Inspired by their initial success, the team ventured into the research and development on VR, augmented reality (AR), serious games and simulation technology for

Y. Cai (✉)
Nanyang Technological University, Singapore 639798, Singapore
e-mail: myycai@ntu.edu.sg

J. Chen
VARTEL Network, Singapore 637144, Singapore

Y. Cai and Q. Cao (eds.), *When VR Serious Games Meet Special Needs Education*,
Gaming Media and Social Effects, https://doi.org/10.1007/978-981-33-6942-9_1

special needs education. A good number of VR and serious games have been developed since then to assist children with special needs in their academic, emotional and skill learning.

Singapore like many other countries has been experiencing huge difficulties since early 2020 due to the COVID-19 pandemic. This book was initiated at the beginning of the circuit breaker lockdown in response to the COVID-19 pandemic and completed by the end of work from home in September 2020. It is a collection of project developments on VR and serious games for special needs education.

The remaining parts of the book are organised as follows: Chapter "Learning to Take a Shower Through VR Serious Gaming" discusses a serious game developed for children with ASD to learn shower taking. Chapter "Design of a Virtual Home for Special Needs Children to Learn Life Skills" presents a serious game for children with ASD to learn life skills at home. Chapter "Learning to Cross Roads Through VR Playing" describes a VR game for children with special needs to learn road crossing. Chapter "Virtual Pink Dolphins and Lagoon" illustrates the design of Virtual Pink Dolphins and Lagoon for special needs education. Chapter "Serious Game Design for Virtual Dolphin-Assisted Learning" details the serious game of virtual pink dolphin. Chapter "Evaluation of Serious Games for Special Needs Education" studies the evaluation of serious games for special needs education. Chapter "Game-Assisted Vocational Training" devotes to the serious game design for vocational training. Chapter "Design of a Home Bag-Packing Serious Game for Children with ASD" shares the Design of Home Bag-Packing a serious game designed for children with ASD to learn planning skill. Chapter "iPad Serious Game to Aid Children with Special Needs in Emotion Learning" explores the design of an iPAD serious game for emotion learning in special needs education. Chapter "Design of a VR Supermarket Serious Game" examines the design of a VR serious game for children with special needs to learn shopping in supermarket.

Acknowledgements This book will not be possible without the supports from Nanyang Technological University, AWWA School, METTA School, Suzhou Industrial Park Renai School, Singapore Millennium Foundation, Institute for Media Innovation, Institute for Mental Health and many individuals who render their help one way or another.

References

1. Cai, Y., Lu, B.F., Fan, Z.W., Chan, C.W., Lim, K.T., Qi, L., Li, L.: Protein immersive games and computer music. Leonardo **39**(2), 135–138 (2006). https://doi.org/10.1162/leon.2006.39.2.135
2. Cai, Y., Chia, K., Thalmann, D., Kee, K., Zheng, J., Thalmann, N.: Design and development of a virtual dolphinarium for children with autism. IEEE Trans. Neural Syst. Rehabil. Eng. **21**(2), 208–217 (2013). https://doi.org/10.1109/TNSRE.2013.2240700
3. Lu, A., Chan, S., Cai, Y., Huang, L., Nay, Z.T., Goei, S.L.: Learning through VR gaming with virtual pink dolphins for children with ASD. Intera. Learn. Environ. **26**(6), 718–729 (2017). https://doi.org/10.1080/10494820.2017.1399149

Learning to Take a Shower Through VR Serious Gaming

Song Eun Matthew Teoh, Qi Cao, and Yiyu Cai

Abstract This chapter describes an interactive virtual reality (VR) serious game for children with autism spectrum disorder (ASD) to learn life skills. The work is a result from a collaboration between Nanyang Technological University and a special needs school in Singapore. Teaching children with ASD some tasks of life skills may be challenging. For example, learning to take a shower can be a struggle for children with ASD as they have difficulty understanding the concept of hygiene. This work uses the Microsoft Kinect to provide players with an engaging experience, while teaching them the steps to take a shower in a VR environment. The experiment has been conducted with six boys around the age of 8 from the special needs school. It is observed the experiment is very successful, eliciting excitement and interactivity for the students to learn life skills such as shower taking.

Keywords VR gaming · Life skills · Children with ASD · Shower taking

1 Introduction

1.1 Autism Spectrum Disorder and Teaching Aids

Autism is a lifelong neurodevelopmental disorder, which affects the brain and psychic-emotional system. Those affected usually have difficulty with emotional expression and recognition, difficulty with social relationships, difficulty with communication skills, etc. Children with autism spectrum disorder (ASD) may exhibit delayed or abnormal language development, and tend to be preoccupied

S. E. M. Teoh · Y. Cai (✉)
School of Mechanical and Aerospace Engineering, Nanyang Technological University, Singapore 639798, Singapore
e-mail: myycai@ntu.edu.sg

Q. Cao
School of Computing Science, University of Glasgow, Singapore Campus, Singapore 567739, Singapore
e-mail: qi.cao@glasgow.ac.uk

Y. Cai and Q. Cao (eds.), *When VR Serious Games Meet Special Needs Education*,
Gaming Media and Social Effects, https://doi.org/10.1007/978-981-33-6942-9_2

with repetitive, stereotyped behaviors or interests. They may also exhibit a general resistance to changes in routine [1].

The causes of autism are unknown. There is such a wide range of symptoms and severity, leading to the use of ASD as a means to group together these diagnoses, such as autism, Asperger syndrome, childhood disintegrative disorder, and pervasive developmental disorder. Additionally, certain percentage of children with ASD are also mentally disabled [2].

There is currently no known cure—only treatments are available, with the most well-known being intensive behavior modification therapy. Other alternative treatments are play therapy, music therapy, occupational therapy, social skills training and, sometimes, prescribed drugs. Often, a combination of these treatments is used.

Many digital teaching aids have been developed for children with ASD to learn various types of skills from communication and psychomotor training, to social behavior augmentation [3]. Social storytelling could be helpful for children with ASD to better understand social situations [4]. These stories are used to explain emotions and interpersonal relations. Storytelling is an engaging approach for teaching.

Children with ASD tend to have difficulty coping with unstructured time [5]. The use of visual schedules helps these individuals remember what is going to happen and the order in which these events will occur. This approach can also be applied to checklists, which helps to break down a complicated task.

Play therapy is also often used to help grow and develop their minds. It provides a dynamic, enthusiastic, and play-oriented approach to work with children as well as facilitates socialization and relationship building [6].

Zakari et al. [7] have reviewed about 40 serious games with various learning objectives. These serious games are in either 2D or 3D graphics and running on different computational platforms. A total of 43 serious games for people with ASD or intellectual disabilities (ID) have been reviewed by Tsikinas et al. [8]. The possibility is explored for these serious games to provide blended learning with formal and informal education to people with ASD or ID.

1.2 Virtual Reality and Hygiene Education

Immersive and interactive learning is increasingly recognized as a ubiquitous form of learning today [9, 10]. The use of virtual environments and VR serious games to teach real-world skills to children with ASD is a growing trend. VR environments are helpful for children with ASD to learn skills for independence. Children with ASD are encouraged to try these out in the real world and gain real-life experience [11].

Serious games act as simulations of real-world events. Serious games have been adopted in various applications such as health care, rehabilitation, education, and training[8, 12]. As knowledge acquisition tools, digital games for learning are developed across varied topics in Science, Technology, Engineering, and Mathematics

(STEM) subjects, as well as health education [13]. Molna and Kostkova [14] present a serious game for children that focuses on creating awareness about hygiene and responsible antibiotic use. A VR hand washing game for preschool and lower primary school children is introduced to teach keeping good hand hygiene [15]. An augmented reality serious game for health care worker's hand hygiene education is presented by Klinker et al. [16]. A browser-based educational game is depicted that aims to improve hand and respiratory hygiene for young people within 9–15 years old [17].

Hygiene is a challenging concept for children with ASD. Often, these children experience a breakdown in understanding the importance of being clean. To teach children with ASD that taking shower is very important to maintain good hygiene. Germs may make children sick if they do not shower to keep good hygiene. Children with ASD also need to be reminded that it is smelly if not taking shower for days. With bad hygiene, it is difficult for children to make friends [18]. There are many overwhelming stimulations when taking a bath or shower—the sound of running water in an enclosed area, the water on the skin, the smells of soap and shampoo, and so on. This can result in a sensory overload, which makes bath an uncomfortable and stressful activity.

Taking care of one's personal hygiene is a constant task. It is also a personal activity, such as going to the toilet and taking a shower. However, these tasks can be challenging for individuals with ASD. They are less likely to be independent in this task, and thus end up reliant on parents and caretakers. This chapter proposes to develop a VR serious game to teach children with ASD how to take a shower. The game story, demo model and game model will be presented in detail next.

1.3 Scope and Objectives of This Research

This work explores the use of VR in the teaching of life skills to children with ASD. The Microsoft Kinect is used in conjunction with the Unity3D game engine to develop an interactive serious game. The objectives of the work are shown as follows:

(1) Design and develop a VR serious game to teach a life skill to children with ASD.
(2) Use the Microsoft Kinect to deliver an immersive and engaging experience.
(3) Gather feedback of teachers and staff from the special needs school.
(4) Conduct testing of the VR serious game for children with ASD.
(5) Evaluate the children's responses and discuss the results of the experiments.

1.4 Organizations of the Chapter

The remaining parts of the chapter are organized as follows: Sect. 2 describes the design of the proposed game. The implementation of the serious game is introduced in Sect. 3. Section 4 presents the design and storyline of the game engine. The experiment results are evaluated and discussed in Sect. 5. This is followed by a conclusion in Sect. 6.

2 Design of the Serious Game

This work explores helping individuals with ASD to develop some independence in taking a shower.

The entire activities have been carried out in three stages as follows:

(1) *Stage 1: Game Story*—A game story has been used to introduce the in-game character, and why the character is taking a shower. The story reinforces the reasons why people need to take showers. It also introduces the symbols, which are seen in the game. The story encourages children with ASD to use the gestures, which they learn in playing the serious game.
(2) *Stage 2: Demo*—Children from the special needs school have the chance to interact with the in-game character based on the Unity Asset Store. They also learn how to interact with the virtual objects in the serious game.
(3) *Stage 3: Game*—Children play the serious game through the steps of taking a shower. There are varying difficulty levels in the serious game. Players can challenge themselves and test their understanding of the activities at different levels.

2.1 Interactive Control of the Serious Game

In literature, Microsoft Kinect, has been used with the Flexible Action and Articulated Skeleton Toolkit (FAAST), is developed by the Institute for Creative Technologies, University of Southern California [19]. The FAAST takes players' gestures as inputs and maps these inputs to outputs such as keyboard commands or mouse clicks. The FAAST allows developers to easily integrate gesture controls with Microsoft Kinect.

However, using gestures may not be the most effective solution when working with individuals with ASD. They may not be able to understand why the gestures are required for the activities. Additionally, depending on the severity of their diagnosis, individuals with ASD may also lack the required coordination to perform the gestures accurately. It may result in their gestures not being detected properly. Furthermore, in addition to learning how to play the serious game, players would need to learn how to use the right gestures to elicit the appropriate control. This could be potential obstacles for them.

This work therefore seeks to explore if a more natural solution is possible, such as using players' body motion as the interaction controller for the serious game.

2.2 Steps of Taking a Shower in the Game Scripts

The serious game focuses on breaking down the showering activities into a series of steps for players as follows:

(1) *Turn on the shower*—First, turn on the shower and wet the skin and hair.
(2) *Turn off the shower*—When the skin and hair are wet, the shower is turned off.
(3) *Put soap on the body*—Once the shower is turned off, lather soap onto the body.
(4) *Put shampoo on the head*—Then, put shampoo onto the hair and scrub it well.
(5) *Turn on the shower*—Next, the shower can be turned on again.
(6) *Rinse everything off*—Once the water is running, begin rinsing off all the soap and shampoo. Make sure there is no more soap on the body or shampoo in the hair.
(7) *Turn off the shower*—Complete the task by turning off the shower.

2.3 Visual Instructions in the Serious Game

Visual instructions have been provided using a checklist. The checklist not only lists all the steps of taking a shower, but also guide players through the steps by indicating the completed steps and the remaining steps separately. Additionally, words and visual symbols will be used concurrently to better help them.

2.4 Visual Feedback

The visual feedback has been provided in different scenes of the serious game. It provides better help to guide players and enhances their performance in the serious game over the time.

(1) *Checklist*—Visual feedback is firstly provided in the checklist. The serious game checks off steps as they are completed by players. This helps players keep track of which steps have been completed, and which steps are yet to be completed.
(2) *Character*—Players receive constant visual feedback from the in-game character. The character responds to their movements and is thus able to show them where they are "spatially" within the game scene. The character chosen will be appropriately attired for the activity with a towel wrapped around the waist, due

to the training for sensitivity to "private parts" to children with ASD. Furthermore, the character has virtual dirt particles at the start of the serious game. These virtual dirt particles need to be removed at the end of the game. It is used to teach players that taking a shower makes the body clean.

(3) *Shower*—Virtual particles are used to simulate running water in a shower. These virtual particles fall from above the in-game character to give the impression that water is really falling onto the character. Game players can get the visual feedback in the status of the shower being turned on or turned off.
(4) *Soap and Shampoo*—Virtual particles resembling shower foam appear on the character's body and head. This visual representation also serves as a reminder for players that when they take shower, they need to get the soap all over their bodies and shampoo thoroughly in their hair. The virtual soap and shampoo are also of different colors to help them recognize that these items are different.
(5) *Rinsing*—This is a key step in the training. The visual feedback received in this step is the disappearance of the virtual foams of soap and shampoo as players "rinse" their body by scrubbing all over the place. The virtual soap and shampoo particles will be removed. The virtual dirt particles will be removed as well. This reinforces the concept that taking a shower makes you clean.
(6) *Completion*—A virtual arrow will point to each icon in the serious game, as it appears as hints for players to interact with that virtual icon. This is to help them progress through all steps accordingly.
(7) *Scene* —The scene will simply consist of the character against a virtual "wall" textured to look like a bathroom environment. This is to add to the visual image of a character ready to shower. This minimalist approach will aim to reduce the amount of unnecessary visual noise, which could distract players. Finally, the completion of the activities will be visually indicated by virtual fireworks and the appearance of a checkmark symbol.

2.5 Audial Instructions

Besides the visual instructions in the serious game, players are prompted by clear and concise verbal instructions at each step of the game. Both approaches can provide better guidance to them.

2.6 Audial Feedback

There are several game scenes with audial feedback functions.

(1) *Selection*—A "ping" sound is played whenever players touch an icon with their hands. This indicates, to them, that the icon has been activated.

(2) *Shower*—There is sound of running water playing whenever the shower is turned on. This complements the visual feedback of seeing waterfall onto the in-game character, indicating that the shower is on.
(3) *Cheering and Fireworks*—Upon completion of the shower activities, fireworks appear on the screen with the sound complementing that effect. Furthermore, there will be sounds of children cheering, to also signal that the activities have been successfully completed.

2.7 Various Difficulty Levels

The difficulty levels will be varied based on the amount of prompting given to players. The easy level setting has both visual prompts (i.e., checklist and arrow) and audial prompts for verbal instructions. The medium difficulty level does not have verbal instructions. The hard level setting has neither the verbal instructions nor the visual (i.e., checklist and arrow) prompts.

This enables players to try the activities multiple times while gradually making it more challenging. It also helps in evaluating their progression and understanding of their performance.

3 Implementation of the Serious Game

3.1 Introduction to the Microsoft Kinect

The Microsoft Kinect is a motion-sensing input device developed by Microsoft, which is shown in Fig. 1. It is based on technology invented by PrimeSense. The Microsoft Kinect, as a 3D depth camera, was initially used for the XBOX game console. It creates various applications for multimedia computing and new ways for playing video games and experiencing entertainment [20].

In February 2012, Microsoft released a software development kit (SDK) for Windows PC, which potentially transforms new means for human–computer interaction [20]. Various software applications can be developed by different parties with the SDK. Incorporating functions such as full-body 3D motion capture, facial recognition, and voice recognition capabilities, Microsoft Kinect enables users to interact

Fig. 1 Microsoft Kinect device

naturally with computers by motion, gesture, or voice. The first generation of dedicated Kinect for Windows unit along with the commercial SDK was released in 2012.

The Microsoft Kinect features an RGB camera, a depth sensor which consists of an infrared laser projector combined with a monochrome CMOS sensor. It has multi-array microphone capabilities. There is also a motorized tilt mechanism in the base, which adjusts to find the floor after automatically examining players's environment.

3.2 Introduction to Unity3D

Unity or Unity3D is a cross-platform game engine with a built-in integrated development environment (IDE) developed by Unity Technologies. For each game project, all game scenes and assets such as textures, models, and sounds are contained in a large folder. A game scene is made up of GameObjects which are the objects such as cubes, lights, models, and cameras. These objects can interact with each other in the game through physics and scripts.

Unity provides an asset store where developers can purchase and sell their game assets. There are also many game assets provided free of charge in the asset store.

3.2.1 Unity Scripting

The Unity3D engine allows users to write behavior scripts in JavaScript or C#. Unity scripts can be attached to GameObjects or work on their own. Additionally, variables declared as public in a script will be visible in the Unity3D inspector, allowing for easy manipulation.

3.2.2 Kinect Interface

The Kinect with MS-SDK published by RF Solutions [21] is available for free of charge in the Unity Asset Store. This Unity asset uses scripts to interact with the Microsoft Kinect. The provided sample scene demonstrates how to create a Kinect-controlled avatar for Unity projects.

A model is required to use the AvatarController script. Developers drag corresponding GameObjects from the model to the dialog boxes in the inspector. The head of the model is being mapped to the corresponding head control as shown in Fig. 2. The Kinect with MS-SDK asset is also able to recognize gestures.

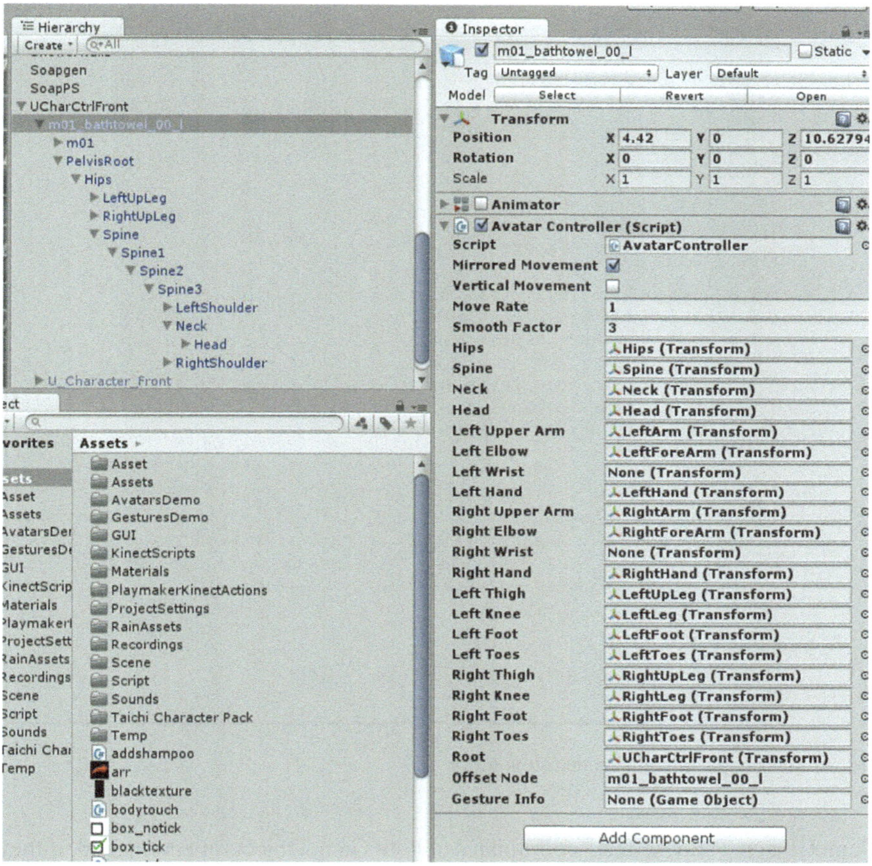

Fig. 2 Mapping the Kinect controller

3.2.3 Collision Detection

Collision detection is to check if a GameObject collides or touches with another GameObject in the interactive VR serious game. Native collision detection function is available as a part of the Unity3D engine environment. However, the native collision detection requires precise actions, which is observed to be a challenging task for players with ASD. Many trials and attempts are made to implement the native collision detection which is unfortunately less effective.

In the end, collision detection is implemented through scripting. The script would compare the coordinates of various GameObjects, such as hand, head, body, and icons. If these GameObjects are within the same area, it is considered a collision being detected.

A sample of collision detection function is shown in Fig. 3. In this example, the script is checking to see if the right hand GameObject collides with the head

```
// Use this for initialization
void Start () {
    s = 0.15f;
    z = 0.1f;
}

void detect(){

    headx = Head.transform.position.x;
    heady = Head.transform.position.y;
    headz = Head.transform.position.z;

    handrx = Hand_r.transform.position.x;
    handry = Hand_r.transform.position.y;
    handrz = Hand_r.transform.position.z;

    handlx = Hand_l.transform.position.x;
    handly = Hand_l.transform.position.y;
    handlz = Hand_l.transform.position.z;

    //test for handr
    if (handrx <= (headx + s) && handrx >= (headx - s)) {
        if (handry <= (heady + s) && handry >= (heady - s)) {
            if (handrz <= (headz + s + z) && handrz >= (headz - s - z)) {
            //If we get here means that there was a collision detected!
                boxr.renderer.material.color = Color.blue;
            }
        }
    } else {
        boxr.renderer.material.color = Color.white;
    }
```

Fig. 3 An example of collision detection script

GameObject. The *x*, *y*, and *z* coordinates of both GameObjects are compared, if they are close to each other within certain threshold values of the sensitivity factors (*s* and *z*) in Fig. 3. The sensitivity factors can be easily modified in the script, catering to various levels of participant motor coordination by different players.

3.2.4 Particle Systems

The Unity3D particle systems engine is used to implement the virtual soap, shampoo, dirt, fireworks, and shower into the VR serious game. There are many options to design particle systems to suit the desired needs, as shown in Fig. 4. Many of these options can be manipulated by scripting, with some applying to particle systems individually and others applying to particle systems as a whole.

These are some particles with important features in the designed VR serious game, shown as follows:

(1) *Particle Lifetime*—It is the number of seconds that a particle will "live" in the game. The particle will disappear when its lifetime is less than or equal to zero second.

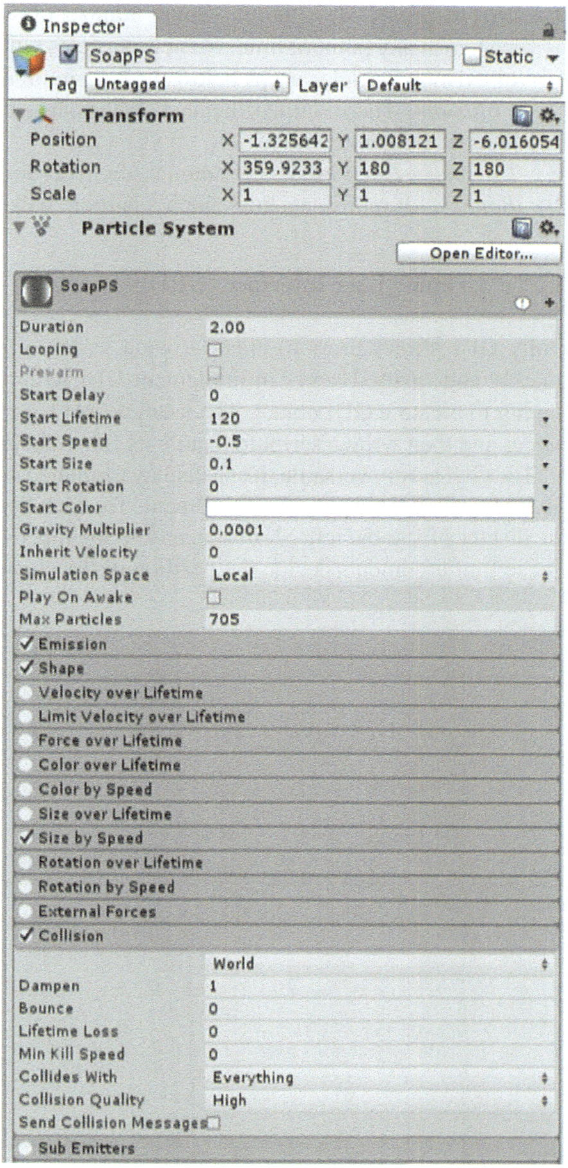

Fig. 4 Particle system interface

(2) *Start Speed*—It is the initial velocity of a particle.
(3) *Start Size*—It is the starting size of a particle.
(4) *Start Color*—It is the starting color of a particle.
(5) *Gravity Multiplier*—It is the effect of gravity on a particle.

(6) *Max Particles*—It is the maximum number of particles allowed to exist in the system at any point of time. No new particles will be emitted unless the number of existing particles drops below the maximum number.
(7) *Collision*—These settings dictate how the particles interact with other GameObjects in the game.
(8) *Dampen*—It represents the amount of energy lost in a collision.
(9) *Bounce*—It represents how much a particle will bounce upon collision.

3.2.5 Graphic User Interface (GUI)

Unity GUI allows users to create a wide variety of highly functional GUIs very quickly and easily. Users can implement GUI with a few lines of code, instead of having to create a GUI object. The users do not have to manually position the GUI object and then write a script to handle its functionality.

The GUI is very versatile. It can display formatted text and images which are called as 2D textures. There are four arguments for each "new Rect", which represent (1) the distance from the left, (2) distance from the top, (3) the width, and (4) the height. These four arguments decide the position of the elements in the GUI.

```
// Checklist - visual prompts for easy and medium difficulty
void OnGUI () {
    if (checklistsignal.renderer.material.color == Color.cyan) {
        GUILayout.Label (blacktexture);

        GUI.Box (new Rect (0,10,260,400), "<size=30>Checklist</size>");

        // step 1
        if (easymode) {
            GUI.Label (new Rect (60,47,40,40), showerimage);
            GUI.Label (new Rect (100,55,120,30), "<size=20>ON</size>");
        } else {
            GUI.Label (new Rect (60, 55, 200, 30), "<size=20>Turn ON SHOWER</size>");
        }

        if (stepnumber <= 1) {
            GUI.Label (new Rect (10, 50, 40, 40), notick);
        }
        if (stepnumber > 1) {
            GUI.Label (new Rect (10, 50, 40, 40), tick);
        }

        // step 2
        if (easymode) {
            GUI.Label (new Rect (60,87,40,40), showerimage);
            GUI.Label (new Rect (100,95,120,30), "<size=20>OFF</size>");
        } else {
            GUI.Label (new Rect (60, 95, 200, 30), "<size=20>Turn OFF SHOWER!</size>");
        }

        if (stepnumber <= 2) {
            GUI.Label (new Rect (10, 90, 40, 40), notick);
        }
        if (stepnumber > 2) {
            GUI.Label (new Rect (10, 90, 40, 40), tick);
        }
```

Fig. 5 GUI script segment implementing the checklist

A segment of code represents the checklist in the game which is shown in Fig. 5. The GUI is called by the function "OnGUI." In this case, it is placed in an IF statement such that the checklist is only visible under certain conditions.

3.2.6 Symbols and Icons

The symbols and icons used in the game are drawn using Microsoft Paint. They are added to the Unity Project Assets and then implemented into the game as 2D textures. They are called in scripts, such as in the checklist. They are also put on GameObjects as interactive symbols/icons. An example of a soap texture added in the game is shown in Fig. 6.

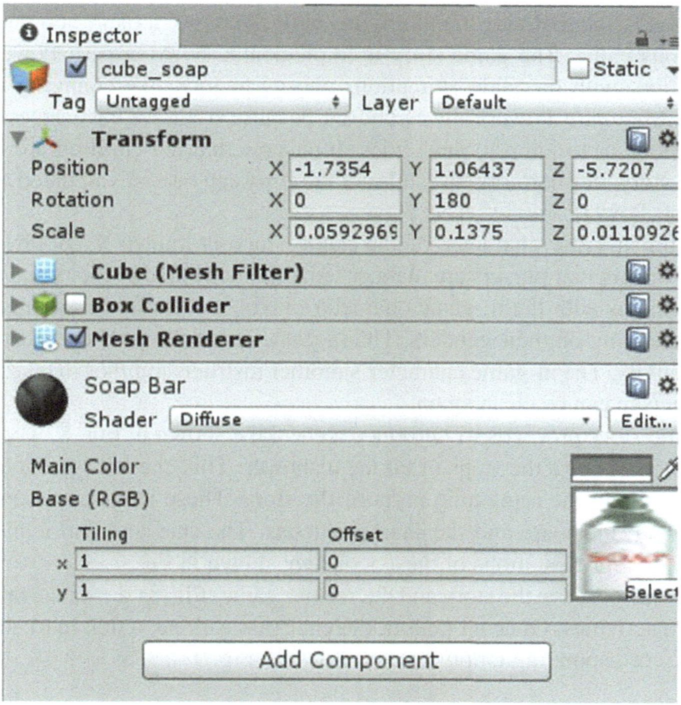

Fig. 6 Soap texture applied to a GameObject

Fig. 7 Examples of in-game icons

4 Game Design

4.1 Game Story

The purpose of the game story is to introduce the game to players by providing context. It gets them slowly familiarizing with the symbols and actions required in the serious game. The game story is implemented by Microsoft PowerPoint as a presentation, with the slide animations chosen to look like changing the pages of a book. Microsoft PowerPoint is chosen because it allows easy access to go to previous slides or to jump to any slides. It ensures that the children are following along with story. Furthermore, the slides of the story can be easily updated and edited in Microsoft PowerPoint.

The main in-game character of the game story is known as Bobby or Judy, depending on whether players are male or female. The character looks androgynous. The story begins with the in-game character playing soccer or gardening, with the activity depending on their genders. The in-game character ends up getting muddy after the activity. The in-game character's mother instructs him/her to take a shower, such that he/she can be clean again.

The game story proceeds to introduce a checklist shown in Fig. 8, where the in-game character knows the steps of taking a shower. This checklist has icons which appear throughout the remaining parts of the story. These icons correspond to the virtual shower, the soap, and the shampoo icons. The start and end icons are also included. The representations of these icons are shown in Fig. 7.

The checklist used in the story and the serious game (Fig. 8) shows the progression through the activities. Once an item in the checklist is done, a tick marked box will replace the corresponding empty box as shown in Fig. 9.

4.2 Demo Mode

There is a demo mode in the serious game. The demo mode is used to link the game story to the serious game. It also serves the purpose to explain to players the methods of how to control and interact with the serious game.

As shown in Fig. 10, the demo mode consists of the in-game character with four of the same icons which can be found in the actual serious game. These items are

Fig. 8 Checklist in game story

Fig. 9 Checkboxes in serious game

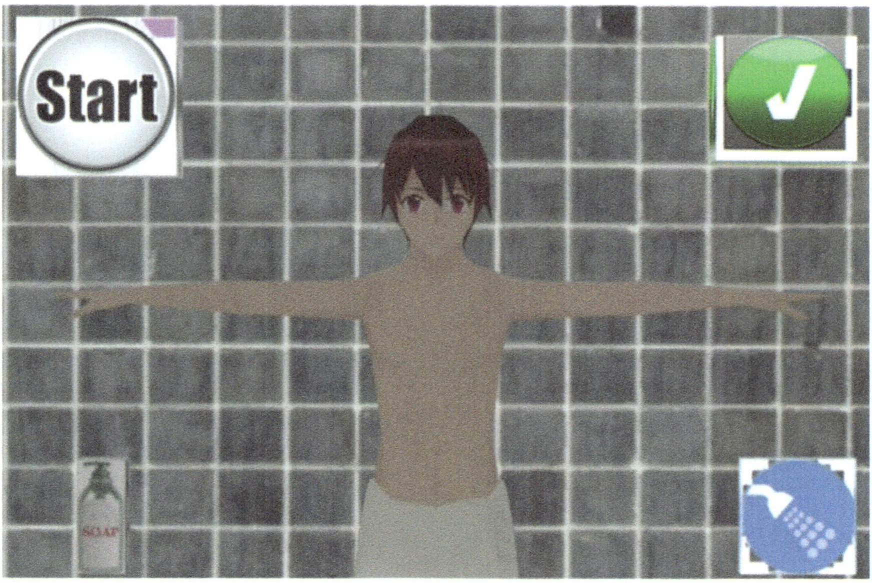

Fig. 10 Screenshot of the demo mode

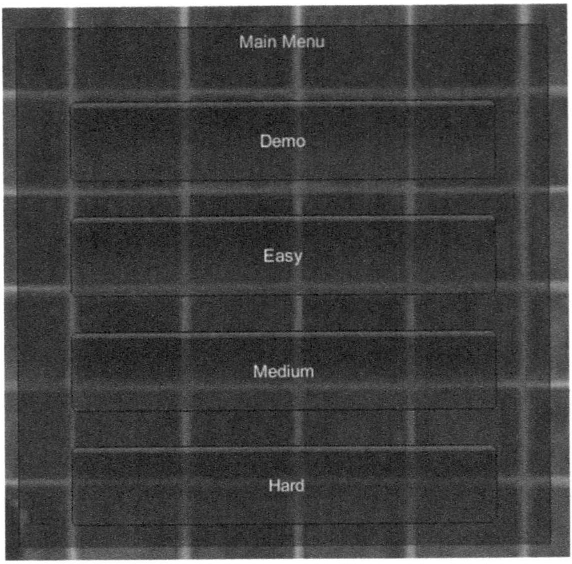

Fig. 11 Main menu of the actual game mode

in the same position as they are found in the actual game. Players will observe how the in-game character responds and moves the body. They are expected to try to get the in-game character to interact with the icons . Particles will be emitted when the virtual objects are touched.

4.3 Actual Game Mode

4.3.1 Main Menu

The actual game mode has three difficulty levels for players to choose. The main menu lists the four available options: demo, and three levels of difficulty (easy, medium, and hard), as shown in Fig. 11. The main menu is written entirely in C# with the GUI functionality provided by Unity.

4.3.2 Checklist in Actual Game Mode

The icons of the in-game checklist in the actual game mode are different according to different game options chosen. The checklist for the easy option is identical to the checklist in the game story. The checklist for the medium option replaces the icons with corresponding text words. It increases the difficulty for players to game. In the hard option, there is no checklist for them. It further increases the difficulty for the game play.

The in-game checklist keeps track of each step of the game activities alone with the game progression. It replaces the empty checkboxes with checkboxes with a tick if each step is completed. The checklist is written entirely in C# script with the GUI functionality provided by Unity.

4.3.3 Steps in Actual Game Mode

The entire sequences of the activities for all three difficulty modes and the demo mode are written in a C# script. The game scene begins with the in-game character being with dirt on the body as shown in Fig. 12. The start icon is also on the screen of the start scene.

The serious game begins when the start icon is "touched" by the in-game character's hand, which is controlled by players. The showering activities in the serious game are implemented in seven steps, following the designed game scripts as described in Sect. 2. These seven steps are described in detail as follows:

(1) *Turn on the shower*—In this step, the shower icon is shown on the game screen (Fig. 13). Verbal instructions are given to players that it is the time to turn on the shower. When the shower icon is "touched," the selection sound effect is played. The shower particle systems start with the shower sound effect. Players are instructed to wet their skin and hair, as shown in Fig. 14. The shower icon becomes hidden now. The checklist will be updated accordingly.

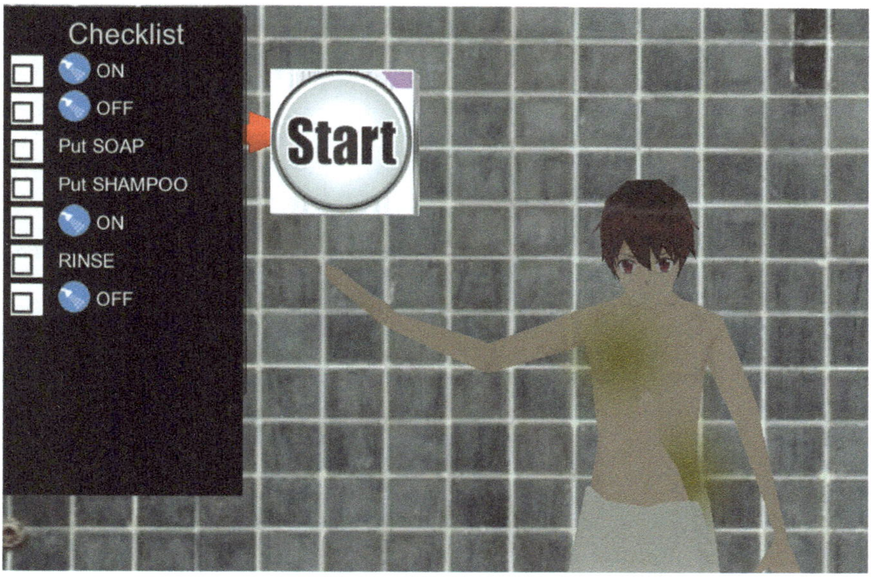

Fig. 12 Start screen of the actual game mode

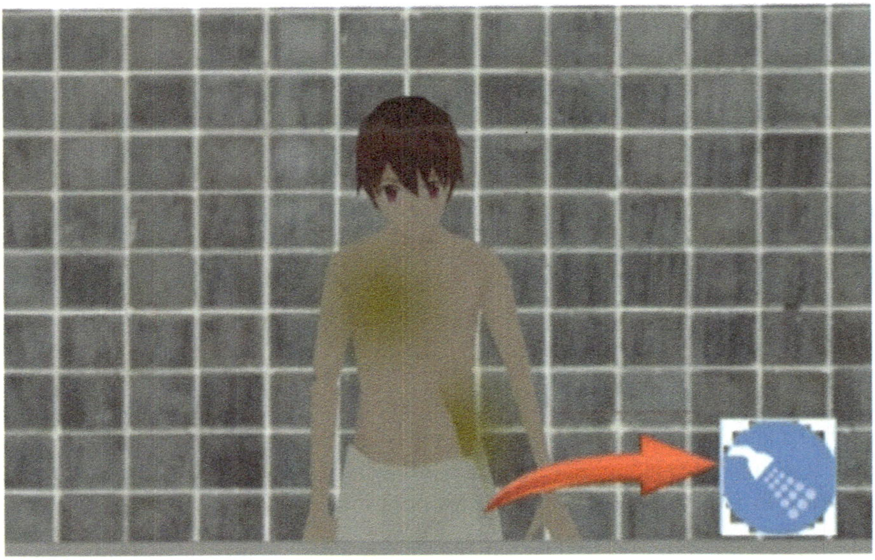

Fig. 13 Step at turn on the shower

Fig. 14 Wet hair and body when the shower is on

(2) *Turn off the shower*—After five seconds, the shower icon is shown again on the screen . Verbal instructions are given to players that it is the time to turn off the shower now. When the shower icon is "touched" with the selection sound effect, the shower particle systems are stopped. The shower sound effect also

stops. It represents the shower has been turned off. The shower icon becomes hidden again. The checklist is updated accordingly again.

(3) *Put soap on the body*—The soap icon is shown on the screen at this step. Verbal instructions are given to players that it is the time to put soap on the body. The selection sound effect plays when the soap icon is "touched." The soap particle systems start, with the soap icon hidden next. Players are instructed to rub the soap all over their body as shown in Fig. 15. The checklist will be updated as well.

(4) *Put shampoo on the head*—At this step, the shampoo icon is shown. Players are verbally instructed to put shampoo on their hair. When the shampoo icon is touched with selection sound effect played, the shampoo particle systems start. Verbal instructions are given to players to rub the shampoo into their hair, as shown in Fig. 16. The shampoo icon will be hidden next, with the checklist being updated.

(5) *Turn on the shower*—The shower icon is shown again in this step. Verbal instructions are given to players to turn on the shower. When the icon is touched, the shower particle systems start with shower sound effect on. The shower icon is hidden, and the checklist is updated (Fig. 17).

(6) *Rinse everything off*—It is the time to rinse everything off at this step. Players are verbally instructed to rinse off all the soap and shampoo. Collision detections are performed among hands, body, and head. If the collisions are detected between character's hands and body, soap particles are removed from the body. If the collisions are detected between hands and head, shampoo particles are removed

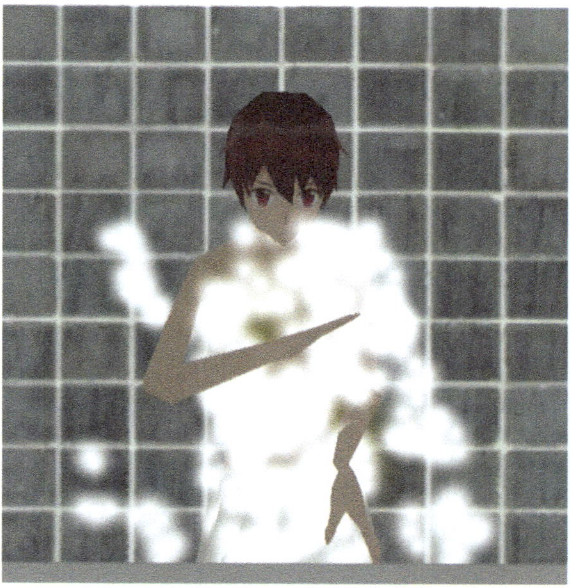

Fig. 15 Put soap on the body

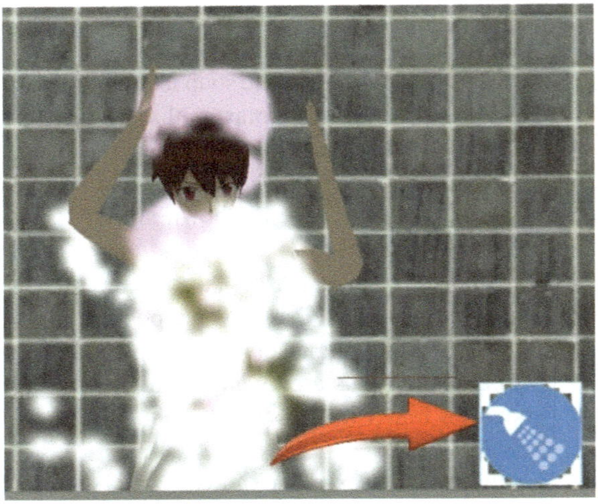

Fig. 16 Put shampoo in the hair

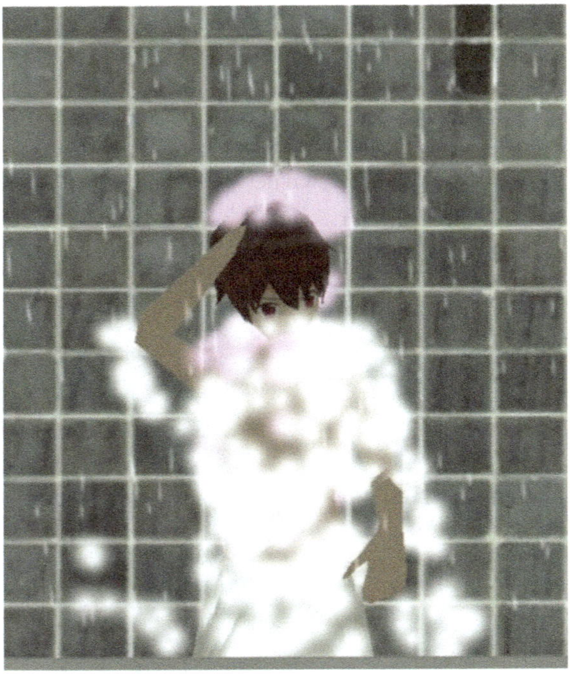

Fig. 17 Rinse everything off

from the hair. When all the particles are removed completely, the checklist is updated.

(7) *Turn off the shower*—At this last step, the shower icon is shown on the screen. Verbal instructions are given to players to turn off the shower. Once the icon

is "touched," the shower particle systems stop with shower sound effect off as well. The shower icon becomes hidden, with the checklist updated accordingly.

Once the seventh step is done, it comes to the end of all activities in the serious game with the end icon displayed. The fireworks particle systems are played. The corresponding fireworks sound effect and cheering are also played concurrently. Next, the game scenes are reset and go back to the main menu (Figs. 18 and 19).

Fig. 18 Time to turn off the shower finally

Fig. 19 Activities completed

5 Experiment

5.1 Experiment Configurations

In this work, the experiment has been carried out in three stages as follows.

(1) *Stage 1: Story*—The story stage is used to introduce the in-game character. It also explains why the in-game character needs to take a shower, as removing dirt and becoming clean. The in-game symbols have been explained throughout the story to help players become familiar with them. The actions and interactions required for the game are also explained to players. They are asked to try out and practice these actions. At the end of the story stage, these in-game symbols and actions have been reviewed to reinforce the meaning to players.
(2) *Stage 2: Demo*—In this stage, players have the chance to interact with the in-game character. They can learn how to interact with the objects in the game. They are instructed to "touch" the various icons to test their understanding of the symbolic meanings. Players can also observe how well of their ability to control the in-game character.
(3) *Stage 3: Game*—In the game stage, players start playing the serious game. It takes them through all steps of taking a shower. They are observed. Their responses and abilities have been evaluated by teachers from the special needs school.

5.2 Experiment Execution

The execution of the experiment has been conducted as follows:

(1) *Location*: Asian Women's Welfare Association (AWWA) School, Singapore.
(2) *Participants*: Six boy students, 8 years old, moderate ASD.
(3) *Equipment*: Laptop with the serious game, Microsoft Kinect for Windows, TV.
(4) *Procedure*: The three stages, i.e., Story, Demo, Game, have been carried out by the participants. The participants have been observed based on their responses to the serious game, as well as their ability to understand the concepts being taught.

The observations and performance evaluations have been conducted by the Head of ICT of AWWA School. Other two teachers from AWWA School have also participated in the observations and provided their input and feedback to the experiment.

According to the three stages of experiment configurations described in Sect. 1, the actual experiment execution takes place as follows.

- *Stage 1: Story*—The game story is told to these children. They are asked to identify the symbols and mimic the actions.

Fig. 20 Experiment–starting the game

- *Stage 2: Demo*—The children take turns to practice and try out the demo mode. They are first instructed to stand on a red mat on the floor, to position themselves in front of the Microsoft Kinect and the TV. They are then instructed to interact with the various icons, with their responses being evaluated.
- *Stage 3: Game*—The children take turns to play the serious game. They stand on the red mat and interact with the game using their hands, as shown in Figs. 20 and 21. Verbal prompts from the teachers are allowed if the children appear to be confused or stuck at certain steps.

5.3 Experiment Results

After the completion of the experiments with all participants, the experiment results have been recorded and evaluations have been performed.

- *Stage* 1: *Story*—The story is quite engaging and sometimes the children are excited to guess what is going to be the next step. The participants enjoy performing the actions together with the teachers. During the revision, the students show their understanding of the symbols by correctly identifying them.
- *Stage* 2: *Demo*—It is observed from the experiment execution that it is easy for participants to learn how to control the in-game character. But one participant is confused by the instruction to "touch" an icon. He walks up to the TV screen to touch where the icon is displayed. Some physical guidance is initially needed to

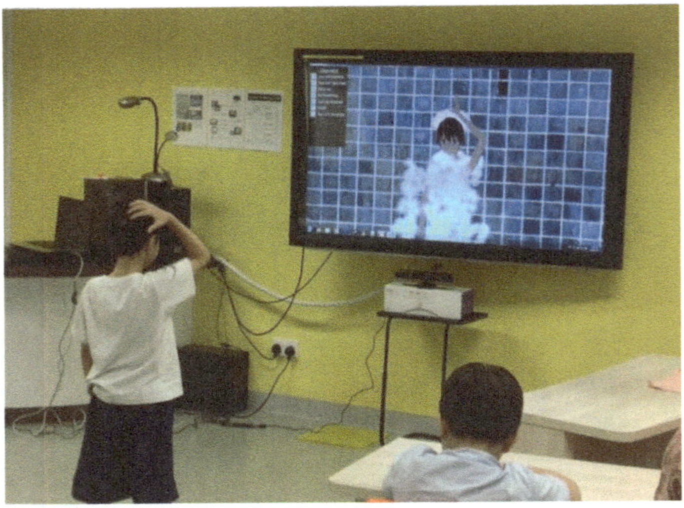

Fig. 21 Experiment execution—showering

another participant who reaches forward to interact with an icon, instead of lifting his arm up by the side.

- *Stage* 3: *Game*—The experiment results show that it is easy for the participants to understand how the serious game is played and interacted. The participants quickly show their learning abilities that they know which the next steps are. When one participant puts the soap in the serious game, he even "put" soap on his back and under his arms. The participants vigorously scrub away the soap and shampoo and enjoy seeing the soap and shampoo disappear in the game. The participants are very excited to see the fireworks and hear the cheering upon the completion of the serious game, and they also cheer along.

The performance of each participant in the *Stage* 2: *Demo* and *Stage* 3: *Game* is recorded and evaluated separately. At the completion of the experiment for all participants, the average evaluation scores for each evaluation item are calculated according to scores of all participants. The average evaluation scores are shown in Fig. 22.

6 Conclusions and Future Improvement

The experiment results show that VR serious game is a useful, exciting, engaging, and enjoyable platform to teach life skills to children with ASD. The experiment results also indicate that the serious game can be simple but effective.

The experiment results show that it is convenient to control in-game characters by using actual actions, especially when players are able to see the in-game character

Teacher's Observations of the Students (Average Scores)

Strongly Disagree	Disagree	Neutral	Agree	Strongly Agree
1	2	3	4	5

<u>Demo</u>

1. The student was engaged during the activity.	[4]
2. The student interacted well with the avatar.	[4]
3. The student tried interacting with the icons.	[3.83]
4. The student understood how to control the avatar.	[4]
5. The student enjoyed the demo.	[3.5]

<u>Game</u>

1. The student was engaged during the activity.	[4.67]
2. The student interacted well with the avatar.	[4.67]
3. The student understood the objective of the game.	[4.67]
4. The student was able to follow the sequence of tasks.	[4.67]
5. The student understood the meaning of the checklist.	[4]
6. The student understood the meaning of the shower.	[4.83]
7. The student understood putting soap.	[5]
8. The student understood putting shampoo.	[5]
9. The student understood washing off the soap/shampoo.	[5]
10. The student understood the meaning of the game.	[4.67]
11. The student enjoyed the game.	[5]

Fig. 22 Average evaluation scores of experiment results

mimicking their actions. In this regard, VR serious game can provide a lot of visual and audial feedback to better train players to interact with the in-game character. It could achieve better training and learning outcomes.

The work is extremely meaningful, as the future of special needs education will become increasingly VR enabled. This work receives positive feedback and has been immediately incorporated as a part of the 3D curriculum at AWWA School, Singapore.

Moving forward, the work will be polished visually with better animations and virtual symbols. Better sound effects and visual instructions could also be implemented. Background music could be added.

It may also be helpful to add in few steps before and after a virtual shower, namely undressing, drying off with a towel, and putting on new clothes. These steps can provide better training to children with ASD to learn the entire showering process.

It may be a natural progression to examine other activities which would benefit from the use of controlling a character with one's own body and actions, such as washing the face, wiping the mouth after eating, or even getting dressed in daily routine activities.

Acknowledgements The authors would like to thank the students, teachers, staffs, principal, and parents of AWWA School for their support, help, and feedback in this research work.

References

1. Sevin, J.A., Rieske, R.D., Matson, J.L.: A review of behavioral strategies and support considerations for assisting persons with difficulties transitioning from activity to activity. Rev. J. Autism Dev. Disord. **2**, 329–342 (2015). https://doi.org/10.1007/s40489-015-0056-7
2. Thurm, A., Farmer, C., Salzman, E., Lord, C., Bishop, S.: State of the field: differentiating intellectual disability from autism spectrum disorder. Front Psychiatry **10**, 256 (2019). https://doi.org/10.3389/fpsyt.2019.00526
3. Liu, G.Z., Wu, N.W., Chen, Y.W.: Identifying emerging trends for implementing learning technology in special education: a state-of-the-art review of selected articles published in 2008–2012. Res. Dev. Disabil. **34**(10), 3618–4362 (2013)
4. Cosgrave, G.: Social stories. https://www.educateautism.com/social-stories.html#.U1PswvmSxs4 (2010)
5. Davies, C.: Using visual schedules: a guide for parents. The Reporter **14**(1), 18–22 (2008)
6. Natasha, P., O'Brien, P.: Play therapy-reaching the child with autism. Int. J. Special Educ. **26**(1), 80–87 (2011)
7. Zakari, H.M., Ma, M., Simmons, D.: A review of serious games for children with autism spectrum disorders (ASD). In: International Conference on Serious Games Development and Applications, Berlin, Germany (2014)
8. Tsikinas, S., Xinogalos, S., Satratzemi, M., Kartasidou, L.: Using serious games for promoting blended learning for people with intellectual disabilities and autism: literature versus reality. In: International Conference on Interactive Mobile Communication Technologies and Learning, pp 563–574. Greece (2017)
9. Cai, Y. (ed): 3D Immersive and Interactive Learning, Springer, Berlin (2013).
10. Cai, Y., van Joolingen, W., Walker, Z. (eds.): VR, Simulations and Serious Games for Education. Springer, Berlin (2019)
11. Cai, Y., Chiew, R., Nay, Z.T., Indhumathi, C., Huang, L.: Design and development of VR learning environments for children with ASD. Intera. Learn. Environ. **25**(8), 1098–1109 (2017). https://doi.org/10.1080/10494820.2017.1282877
12. DiPietro, J., Kelemen, A., Liang, Y., Sik-Lanyi, C.: Computer-and robot-assisted therapies to aid social and intellectual functioning of children with autism spectrum disorder. Medicina **54**(440) (2019)
13. Boylea, E.A., Hainey, T., Connolly, T.M., Gray, G., Earp, J., Ott, M., Lim, T., Ninaus, M., Ribeiro, C., Pereira, J.: An update to the systematic literature review of empirical evidence of the impacts and outcomes of computer games and serious games. Comput. Educ. **94**, 178–192 (2016)
14. Molna, A., Kostkova, P.: Learning about hygiene and antibiotic resistance through mobile games. In: International Conference on Digital Health, pp. 95–99. Lyon, France (2018).
15. Shimada, S., Funahashi, K., Ito, K., Tanase, Y., Iwazaki, K.: VR hand hygiene training system that visualizes germs to be washed and removed. In: IEEE Global Conference on Consumer Electronics. Nagoya, Japan (2017)
16. Klinker, K., Fries, V., Wiesche, M., Krcmar, H.: CatCare: designing a serious game to foster hand hygiene compliance in health care facilities. In: International Conference on Design Science Research in Information Systems and Technology. Karlsruhe, Germany (2017)
17. Farrell, D., Kostkova, P., Lecky, D., McNulty, C.: Teaching children hygiene using problem based learning: the story telling approach to games based learning. In: 2nd Workshop on Story-Telling Approach to Educational Games (STEG), Aachen, Germany (2009)

18. Lingle, J.: Teaching hygiene without meltdowns. Autism Educates. https://www.autismeducates.com/2018/11/06/meltdowns/ (2018)
19. Institute for Creative Technologies, University of Southern California: Flexible Action and Articulated Skeleton Toolkit (FAAST). https://projects.ict.usc.edu/mxr/faast/ (2010)
20. Zhang, Z.: Microsoft kinect sensor and its effect. IEEE Multim. **19**(2), 4–10 (2012)
21. Filkov, R.: Kinect v2 examples with MS-SDK and nuitrack SDK. RF Solutions. https://rfilkov.com/2014/08/01/kinect-v2-with-ms-sdk/ (2014)

Design of a Virtual Home for Special Needs Children to Learn Life Skills

Yee Peng Jolene Chung, Qi Cao, and Yiyu Cai

Abstract This research sets out to use first-person interaction in virtual reality (VR) situations to teach life skills to children with Autism Spectral Disorder (ASD). A virtual home environment is designed and developed for children with ASD for life skill learning in a safe yet immersive manner. Teaching autistic children life skills may be challenging as they generally have cognitive limitations and behavioural problems. Hence, tasks that involve a certain level of risk are usually executed with accompaniment from their caregivers. To mould children with ASD to be more self-reliant, it will be good to provide learning in a safe, flexible, and adaptable environment. Supplying a propitious learning platform with a step-by-step process will be helpful in the progress of their learning. This research uses the Leap Motion sensor and the Unity3D game engine to build an engaging experience for the children to learn life skills such as using an alarm, brushing of teeth, and making a cup of drink. Experiments have been conducted in a special needs education school in Singapore. A group of students and teachers from the school have participated in the experiment study. Very good engagement and interaction in the developed serious games have been observed in this study. Feedback from teachers and participants are positive. In general, virtual reality is beneficial in addressing common learning problems of children with ASD due to its repeatability, predictability and clear guidance that induces safer learning situations.

Keywords VR game design · Children with autism spectral disorder · Life skills

---

Y. P. J. Chung · Y. Cai (✉)
School of Mechanical and Aerospace Engineering, Nanyang Technological University, Singapore 639798, Singapore
e-mail: myycai@ntu.edu.sg

Q. Cao
School of Computing Science, University of Glasgow, Singapore Campus, Singapore 567739, Singapore
e-mail: qi.cao@glasgow.ac.uk

Y. Cai and Q. Cao (eds.), *When VR Serious Games Meet Special Needs Education*,
Gaming Media and Social Effects, https://doi.org/10.1007/978-981-33-6942-9_3

1 Introduction

1.1 Background

All parents are concerned about their children's safety, but parents of children with autism spectral disorder (ASD) may have more challenges when it comes to teaching safety and social skills. Children with ASD need more intensive and highly structured teaching than typically developing children while learning basic life skills. Besides the difficulty of imparting safety skills to the children with ASD, their parents also struggle to justify this imbalance in their social skills level. It has always been recognized that social interaction is an area of difficulty for children diagnosed with ASD. This includes social interaction and social communication. Hence, owing to their cognitive disability, children with ASD will benefit from learning in a safe environment that provides structured and repetitive practices to hone their daily life skills.

There is empirical evidence that motion-based touchless games or interactions can help to promote the attention skills for autistic children with low moderate cognitive deficit, low-medium sensory-motor dysfunction, and motor autonomy [1]. Although these children with special needs know what they should have done differently in a certain task, the problem lies in getting them to do it in the moment. Hence, providing them a static and safe environment to learn life skills will be beneficial in aiding and enhancing their learning capability.

It has been reported that many children with ASD are generally overwhelmed in tasks execution and difficulty making sense of social situations [2, 3]. While some children with ASD may have the desire to perform tasks on their own or interact with others, the processes can be overwhelming that they do not know how to begin or handle [4]. The executive functioning skills and social skills need to be sharpened to improve their level of comprehension, interpretation, self-reliance, and social interaction. Many research works have been reported to help children with ASD to train their skills, so that they can relate the knowledge learnt to practical social world with more confidence.

A virtual and interactive environment is one of the approaches to impart essential skills of self-independence for children with ASD, before encouraging them to practice in the real world [5]. Such interactive learning can enhance the learning motivation and efficiency [6]. Virtual environments can be created with virtual reality (VR) technology, where children can practice challenging social interactions repeatedly in a less-anxiety situation [7–9]. VR serious game, integrating the education and learning components, provides a safe virtual training environment for children with ASD.

This research focuses on basic life skills learning for children with ASD. VR and Leap Motion sensing technology have been used in the serious game, which captures users' gestural movement to react to some decision-making situations. The interactive serious game will mimic a usual daily routine of children, which covers life skills like using an alarm, brushing of teeth, and making of cup of drink for breakfast. This is to accustom the children to daily tasks in the virtual home, so as to train them to be less resistant to changes in environments. Children with ASD can be more motivated to apply what they have learnt in the virtual games into their real-world lives.

1.2 Objectives

This work aims to design and develop a virtual home for children with special needs to learn basic life skills. These life skills are daily routine activities, encouraging children with ASD to perform these tasks independently without others' help. The Leap Motion API is used in conjunction with the Unity3D game engine to develop an interactive serious game environment for the children to pick up the necessary life skills.

The objectives of this research are as follows:

(1) Designing and developing a serious game to educate children with ASD life skills,
(2) Using VR and Leap Motion sensor to deliver an immersive and engaging experience,
(3) Conducting testing of the serious game with children with ASD,
(4) Evaluating and assessing the responses from these children,
(5) Gathering feedback from the children and teachers of two special needs schools, and
(6) Discussing improvements to better enhance their learning effectiveness and efficiency.

1.3 Scopes of This Research Work

This research uses VR to simulate the human-like interaction in a first-person view. It allows players to fully immerse in the game characters. The setting for interaction and learning of basic life skills will be limited to a usual home setting to allow the children with ASD to familiarize themselves with the daily activities on a usual day.

1.4 Organisations of the Chapter

The chapter focuses on research and development of the VR serious game for special needs education. The system design, tools and design software utilized in this work are discussed in Sect. 2. The overview of the design flow of the serious game is presented in the Sect. 3. Section 4 introduces the design of the virtual home serious game. The detailed game development and implementation are depicted in Sect. 5.

The experiment results are evaluated and discussed in Sect. 7. Section 8 concludes this chapter.

2 Design Tools and Software

2.1 Leap Motion Sensor

Leap Motion sensor is a device detecting users' hands and gestural movements as a new form of human-computer interaction [10]. It is a hardware sensor invented by San Francisco-based start-up Leap Motion Inc. This small and compact device is designed to sense the position of the user's hands and fingers roughly within a hemisphere of three feet, a volume of eight cubic feet [10].

The Leap Motion sensor is equipped with two cameras and three infra-red light-emitting diodes (LEDs) to monitor hand movements. It connects to the computer via a Universal Serial Bus (USB) cable.

Instead of using the Microsoft Kinect, this research aims to make use of the Leap Motion sensor to detect the player's hand gestures in the serious game. This will simulate a more realistic environment when the scenario is played in a first-person's view. Moreover, the sensors will capture the finger movements and hand gestures through the tracking algorithms which will interpret the 3D data and infer the positions of the fingers. This communication will be done through the Leap Motion control panel and the data will be organized into an object-oriented application programming interface (API) structure, which will tie in with the Leap Motion input, allowing a motion-controlled user experience.

One major advantage of using the Leap Motion sensor is high and accurate level of detail [11]. The API provides a direct mapping of hands and fingers whereas other available 3D sensory input devices, like the Microsoft Kinect, returns the sensory data in a raw format, which must be analysed by the software thereafter [11]. This will help in building client applications to be faster and more efficient with consistently accurate data. However, having said that, the Leap Motion sensor will not be accurately detected when the hands do not lie in line with the sensor's detection zone. When the hand rotates with its plane perpendicular to the flat surface, the detection deteriorates significantly. Hence, it is only able to detect simple gestures and finger movements or positions that are not obscured from the line of sight of the sensor [11].

With that, the Leap Motion sensor can be used to detect hands movement and display an almost immediate and accurate feedback. It can provide motions of the player's hands grabbing objects, such as toothbrush, cups, etc. It can also provide a visual feedback that aids in providing an immersive experience for players. Due to its ability to track minor movements of the fingers that the Microsoft Kinect is unable to do, this research aims to make use of Leap Motion sensor to engage children with ASD in learning experience of serious games.

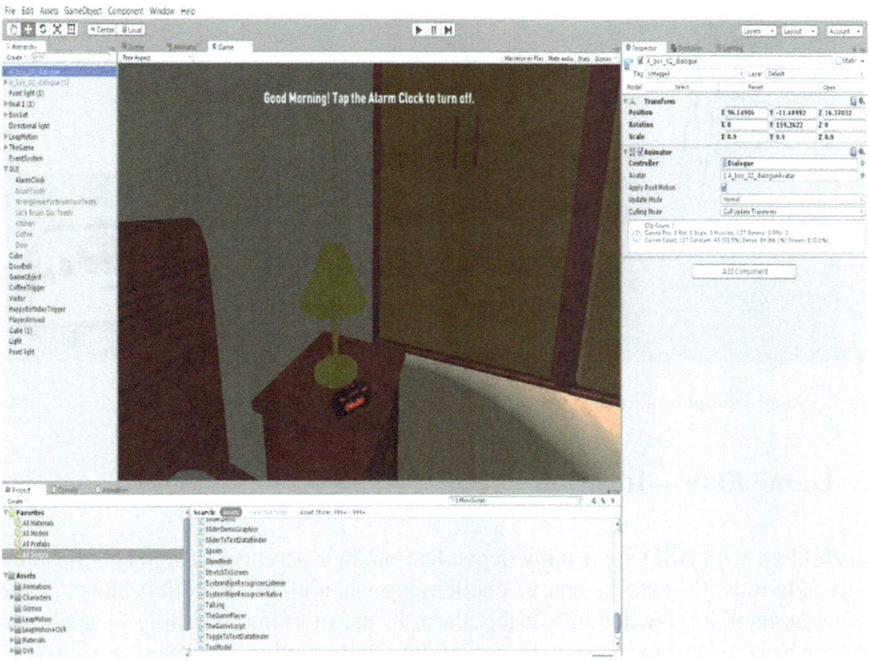

Fig. 1 An example of Unity3D user interface

2.2 *Unity3D*

As a toolset to build digital games, Unity engine is employed in this research. It provides an integrated game development environment with graphics, audio, and physics interaction technologies. With fast prototyping capabilities and large number of publishing targets, Unity3D platform enables rapid games creation and interactive three-dimensional (3D) experiences for players and customers. Each Unity project is a large folder containing all the game scenes and assets which include textures, models, characters, etc. A scene is made up of GameObjects which can interact with each other in the game through scripting and physics. A graphic user interface (GUI) example designed using Unity3D in this work is shown in Fig. 1.

2.3 *Autodesk Maya*

Autodesk Maya is used for rendering, modelling, and simulation. It enables software developers to sculpt and shape models more artistically and intuitively. The platform has good flexibility for efficient user interface design. Some examples of kitchen appliances developed using Autodesk Maya in this work are shown in Fig. 2.

Fig. 2 Some examples of kitchen appliances design

3 Game Flow—Ideation

As children with ASD are usually dependent on their parents or caretakers for their daily activities, these tasks can be challenging when children are left alone. Some activities including switching off the alarm in the morning, brushing of teeth and making a cup of Milo drink may be part of their daily routine. This work aims to help children with ASD develop some independence in performing the aforementioned three activities.

The activities in the serious game flow of virtual home are carried out in three stages as follows:

- Stage 1: Introduction to the virtual home
  A tour around the virtual home will be conducted to allow children to get familiar with the orientation of the home. The virtual home is designed to look akin to the layout of a modern Singapore public housing apartment. Majority of the residential developments of Singapore public housing are high-rise apartment blocks. It is used to enhance the learning capability of the children. The objects in the virtual home serious game will be introduced to them.
- Stage 2: Trial and Demonstration
  Children will have a trial-and-error session by interacting with virtual objects in the virtual home through the Leap Motion sensor.
- Stage 3: Game Playing
  Children will start playing the serious game. It begins with a main menu for them to select their desired activity. Amongst the three activities to be chosen, corresponding buttons are created for them to select. 'Tick' and 'Cross' will be shown to indicate the correctness of their response to the guided hint provided. When the children have tapped onto the correct item, a corresponding button will appear to represent the checklist of the step-by-step process.

3.1 Steps of Game Script

The serious game focuses on breaking down the game activities into several steps. The adaptation of sequencing helps organise children's thoughts more efficiently and, hence, enhance their learning capability.

Game Scene 1: Turning off Alarm Clock.
In this game scene, gamers are required to react to the event of ringing alarm clock. In the morning, the alarm clock rings, until they tap on the virtual button to switch it off.

Game Scene 2: Brushing of Teeth.
In this game scene, gamers are required to tap on the virtual items accordingly in a sequential manner as follows:

(1) Toothbrush—firstly, pick up the toothbrush,
(2) Toothpaste—next, squeeze the toothpaste onto the toothbrush,
(3) Water tap—then, turn on the water tap,
(4) Cup—lastly, use the cup to collect the water from the running tap for rinsing of mouth.

Game Scene 3: Making of Milo Drink.
As Milo is a common local beverage in Singapore, gamers learn the steps in making a cup of Milo drink in the serious game. They are required to tap on the virtual items accordingly and sequentially as follows:

(1) Cup—firstly, pick the cup to be used for drinking,
(2) Milo Powder—next, reach out for the Milo tin to scoop Milo powder,
(3) Hot Water Dispenser—then, proceed to the hot water dispenser to release the right volume of hot water
(4) Spoon—lastly, pick up a teaspoon to stir the mixture.

3.2 Visual Instruction

In the virtual home serious game, visual instructions are provided by incorporating GUI and a checklist to facilitate learning process. The instruction texts guide gamers to pick up the right virtual item in game scenes. The checklist will also appear to indicate the success of action with texts and symbols used concurrently.

3.3 Visual Feedback

With the help of the Leap Motion sensor, a modelled hand is captured on the screen for gamers to interact with. The virtual hand moves accordingly to where gamers

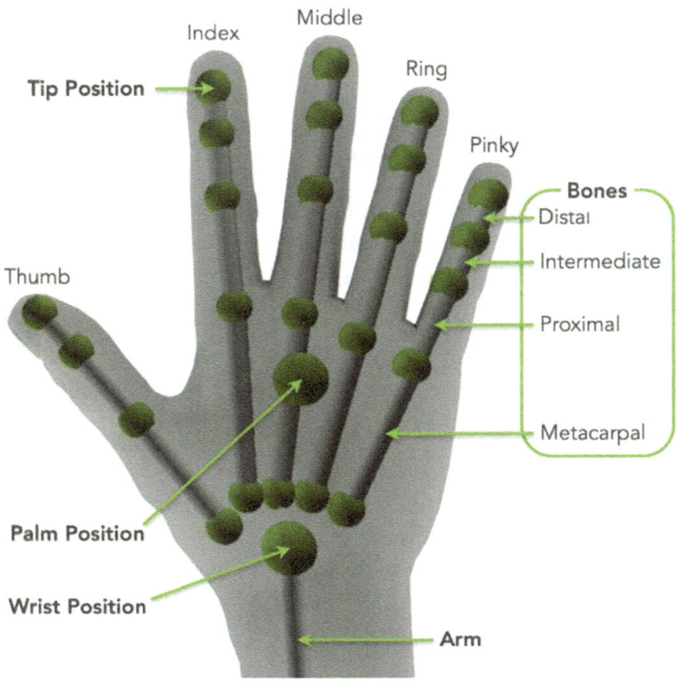

Fig. 3 Leap motion API's positional tracking hierarchy

move hands physically. Leap Motion sensor is able to detect physical hands as it maintains an inner model of a human hand (Fig. 3). It validates the data from the sensor against the hand model. The software can track the positions of the fingers and bones accurately and, thereafter, display it as a visual feedback on the screen.

Assessment of Selection

The assessment is carried out by providing a visual feedback of a 'tick' or 'cross' to indicate a correct or wrong selection, respectively. Due to the target audience of this work, the display of symbols is chosen to exaggerate the selection assessment, as shown in Fig. 4. It can also capture the attention of gamers, by providing a friendly learning environment.

Checklist

A checklist will be shown at the top right corner of every game scene in the serious game. Once a correct virtual item is selected, the name of the item will appear on the right corner of the screen in a sequential order. At the end of the serious game, gamers get to review game steps of the activities they have been engaged in. The learning adaptability is trained as the top-down arrangement aims in providing a hierarchical structure relating to the order of precedence.

Fig. 4 Visual feedback for correct and wrong selections of item

Game Scenes

The games scenes for every activity will display their respective environments that are designed to resemble scenes in real-life situation. Orientation of objects is spaced-out and aligned in the real-world perspective to allow gamers better relating to such environments. Unnecessary objects are left out of game scenes to avoid divided attentions of gamers.

Completion

A virtual button that allows collision detection returns game scenes to the main menu once each game scene has been completed. This is to allow gamers to select other activities after completion of a game scene. Visual feedback is given to guide them to select the virtual 'main menu' button.

3.4 Audial Instruction

With reference to the game scene of '*waking up in the morning*', an alarm ringing sound clip is played, until the player makes a correct selection. This is to impart the idea that the alarm will not go off, until the player taps onto the button of the alarm clock to turn it off. The sound aids in directing them to the correct virtual object.

3.5 Audial Feedback

Upon completion of game scenes of teeth brushing and Milo drink making, an audial feedback will be given to indicate their successions of the tasks. The sound clip for teeth brushing represents the sound of bristles brushing against teeth. The sound clip for Milo drink making represents the sound of a metal spoon stirring and colliding against a cup.

4 Game Design—Virtual Home

In this work, the virtual home is designed similar to the layout of a modern Singapore public housing apartment, which is designed using Autodesk Maya software. The layout of the virtual home is shown in Fig. 5. It tries to create a similar virtual living environment to children with ASD. It brings convenience to children to learn life skills in their familiar environments.

4.1 *Bedroom*

The design of the bedroom inside the virtual home is simply made in Autodesk Maya platform, shown in Fig. 6. The textures of the cupboard, door and carpet are imported and applied onto the objects. Since the scenario starts when the alarm clock rings, an alarm clock is rendered and designed with a snooze button for the gamer to locate. The first game scene, turning off the alarm clock will occur in the bedroom environment.

4.2 *Toilet*

The virtual toilet (Fig. 7) will be used for the step-by-step process of the second game scene, brushing teeth in the serious game. The models of toothbrush, toothpaste, tap,

Fig. 5 Layout of virtual home design for children to learn life skills

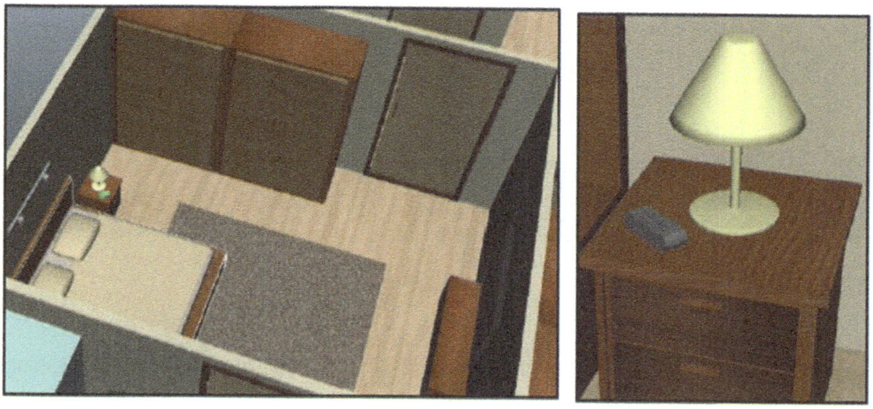

Fig. 6 Design of virtual bedroom with an alarm

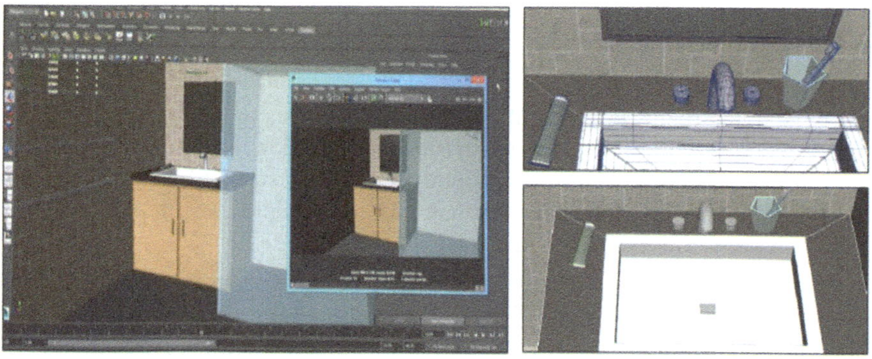

Fig. 7 Design of virtual toilet

and cup are necessary. Moreover, a sink is required for the teeth-brushing game scene. The design of this virtual toilet is to simulate a modern looking toilet in a Singapore public housing apartment. It is to better enhance the players' learning ability.

4.3 *Kitchen*

The kitchen in the virtual home is shown in Fig. 8. For the third game scene of Milo drink making, the necessary virtual objects to be drawn are a cup, Milo powder, spoon, and tumbler (i.e. a hot water dispenser) in the kitchen of virtual home. Other virtual objects like a cake and tray are modelled in the kitchen environment in the game scene. The virtual objects are placed far apart from each other, so that the motion detection and proximity from the Leap Motion sensor will allow more spaces.

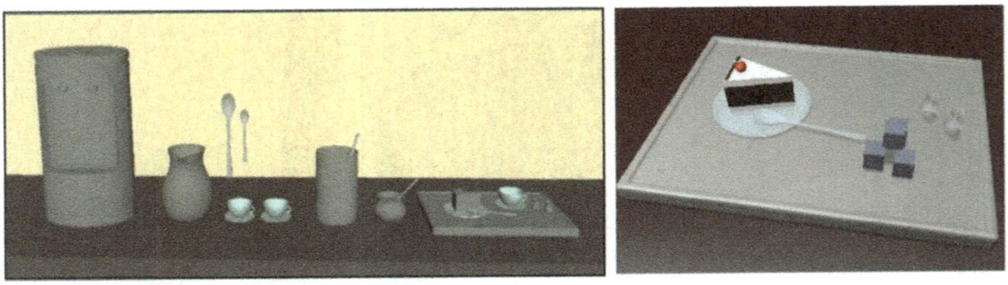

Fig. 8 Design of kitchen appliances

Due to the miniature buttons on the virtual objects, the Leap Motion sensor is not able to capture the extremely accurate and precise hand gesture. Thus, the gestures required in the serious game are only touching or tapping.

4.4 *Living Room*

Since the living room (Fig. 9) will be used to host guests in the real world, virtual objects like sofa, coffee table and a television are created in the virtual home serious game. The area and placement of furniture are akin to a standard living room. This invokes familiarity and aids in the learning process for children with ASD.

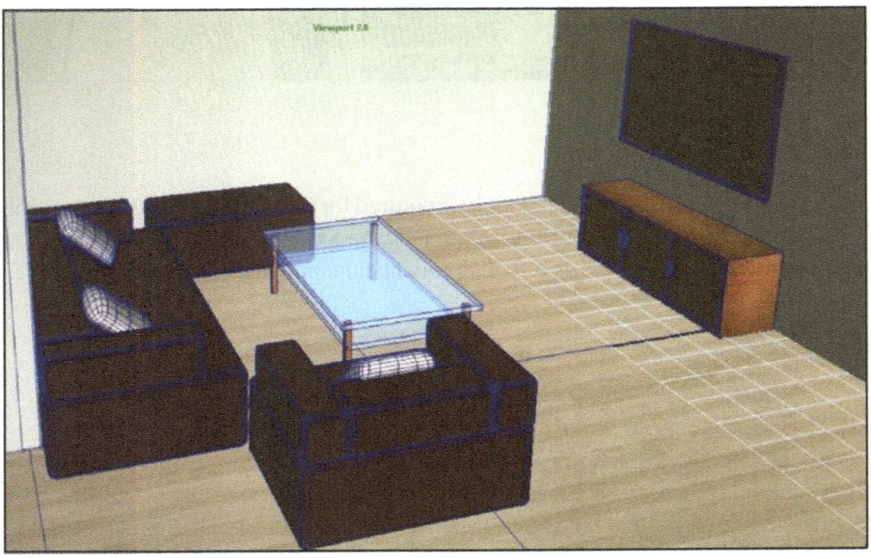

Fig. 9 Design of the living room

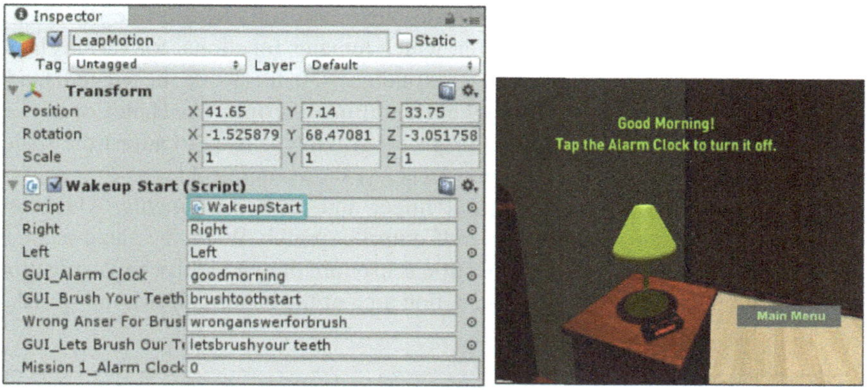

Fig. 10 Script component in inspector window (left) and a game scene GUI (right)

5 Serious Game Development Using Unity3D

5.1 Game Scripting

The Unity3D game engine allows users to write behaviour scripts in Javascript or C#. The MonoDevelop environment is used for compiling and debugging in Unity3D. These scripts will either be attached to GameObjects or they can run on their own. For easy manipulation, variables declared as public in a script are visible in the Unity3D inspector window (Fig. 10).

After scripting is done for task required, MonoDevelop environment allows direct editing and saving of scripts, which can be dragged into the Inspector Window in Unity3D. The serious game will run the scripts when initiated in game mode.

The scripts are used to create graphical effects or control the physical behaviour of objects. In this work, C# programming language is used for scripting. Although Unity uses an implementation of the standard Mono runtime for scripting, objects created in the Unity editor can also be controlled from scripts. Figure 10 shows a game scene for a waking up in the morning.

5.2 Leap Motion SDK

There are modelled hand prefabs assets readily available in the Leap Motion core assets for Unity. For example, HandController is a Unity MonoBehavior instance that serves as the interface between the Unity application and the Leap Motion service. The HandController class will detect and apply the tracking data to the players' hand. The HandModel and FingerModel classes are functions that calls for animating the parts of the hands virtually.

Under the HandModel and FingerModel, there are sub-classes, like the SkeletalHand/Finger and RiggedHand/Finger, which remains functional for various hand designs with corresponding basic structures of objects. In this work, the RiggedHands type is used in conjunction with the RiggedHand/Finger scripts. The HandController-SandBox component in the Unity3D Inspector window is used to adjust the scale, rotation, and position of the hand in the scene, as shown in Fig. 11.

The Leap Motion sensor can detect finger and palm movements, which are controlled by scripts inserted in the HandController prefab. When the game scene is played, hands detected by the Leap Motion sensor are drawn relative to the position and orientation of the prefab. The interaction area of the hand is shown in Fig. 12.

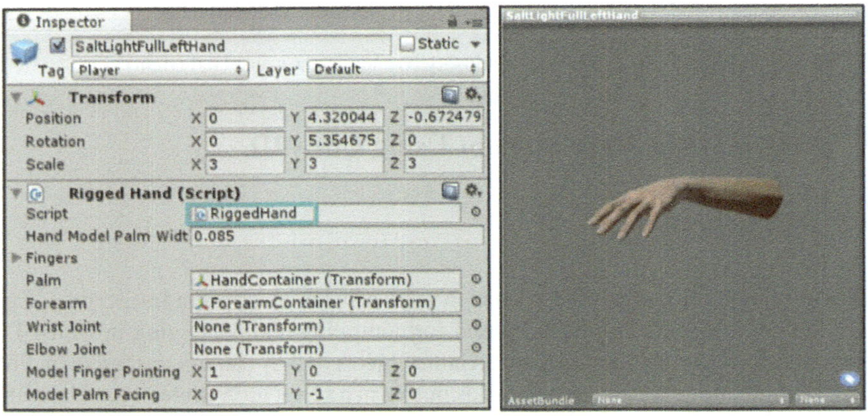

Fig. 11 RiggedHand script in Inspector window (left) and model of a left hand (right)

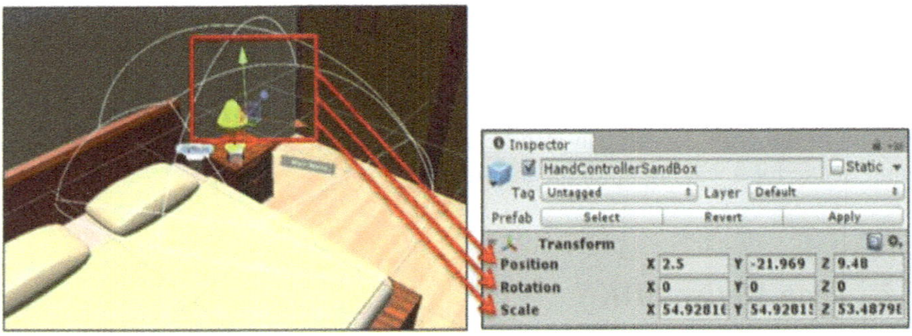

Fig. 12 Dome-like structure for interaction area of the hand

5.3 Collider

Collision detection is made available as a part of the Unity3D environment. The collision detection is also done through scripting. Upon triggering of the corresponding selection, the option is entered. Box Collider is added as a component in the inspector window to allow the selection area to be defined. The Box Collider is used frequently in the game when the buttons are triggered by collision of the hand detected by the Leap Motion sensor and the area specified in the Box Collider.

5.4 Graphical User Interface

Unity3D game engine allows developers to create highly functional GUIs efficiently. The Rect Transform component is the 2D layout representing a rectangle that a user interface (UI) element can be placed inside. The Rect Transform helps to position the component for any UI element within the canvas and is also used for position adjustment. GUI can display text and images with meshing 2D textures. An example of UI image is shown in Fig. 13.

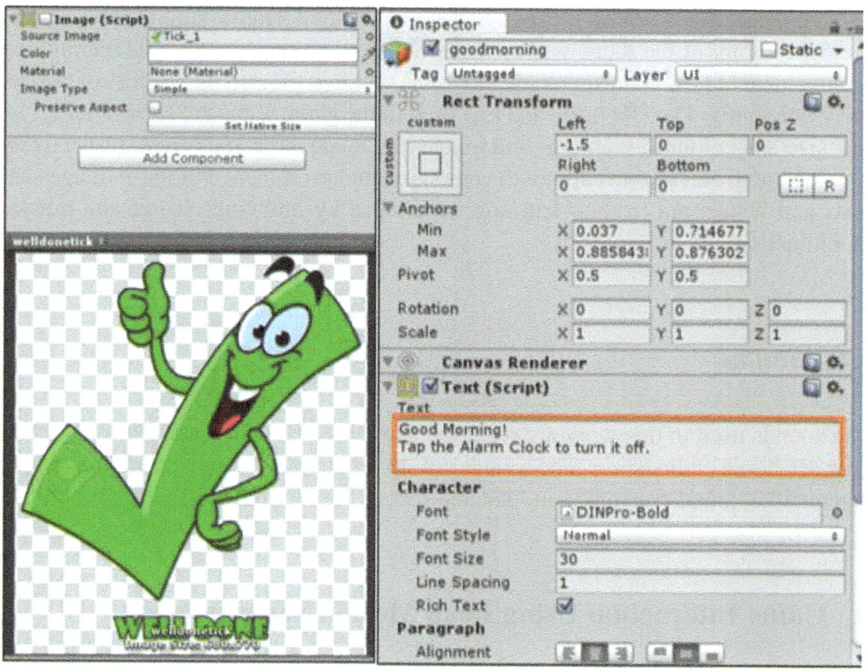

Fig. 13 UI image for correct choice and UI Text for waking up in the morning scene

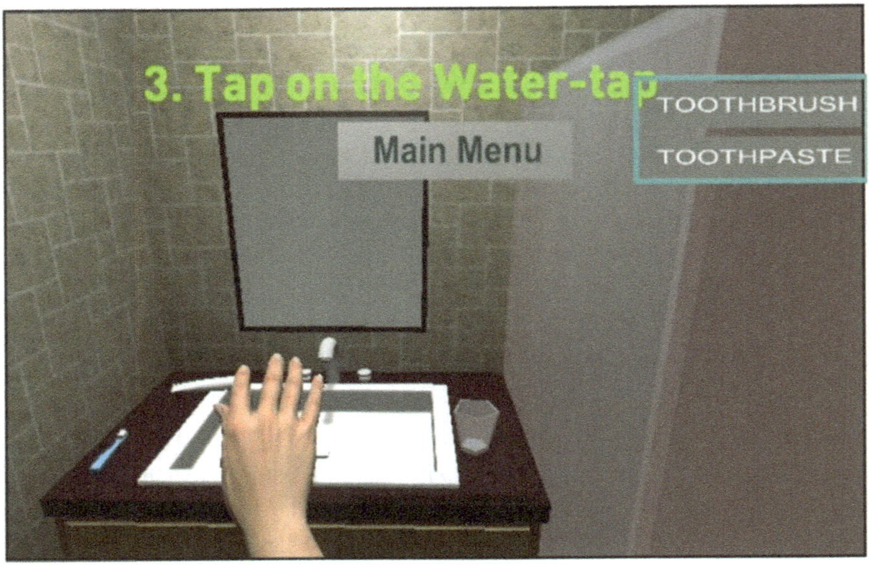

Fig. 14 UI button with text for completion of steps in Teeth Brushing scene

To enhance the learning process, a checklist in the form of buttons appearing on the right corner of the scene will appear when a selection is made. The hierarchical arrangement of top-down approach allows the children to learn the process of the task more efficiently. The UI button and text used in the game scene are shown in Fig. 14. The size and colour of wordings and images are made large and bright, respectively. This will help children to capture the emphasis and prominence of these images and texts and hence, aid in their learning experience by allowing clearer and quicker identification.

5.5 Sound

The sounds used in this work are recorded and edited using TwistedWave Online, a browser-based audio editor. The sounds are stored in SoundCloud, an audio platform that enables sound creators to upload and record the original sound clip.

6 Game Interaction Using Leap Motion Sensor

The game is implemented in the first-person perspective to provide an immersive environment that mimics the real-life situation for children with ASD to learn more effectively. Prior to the starting of the game, a demonstration and introductory session

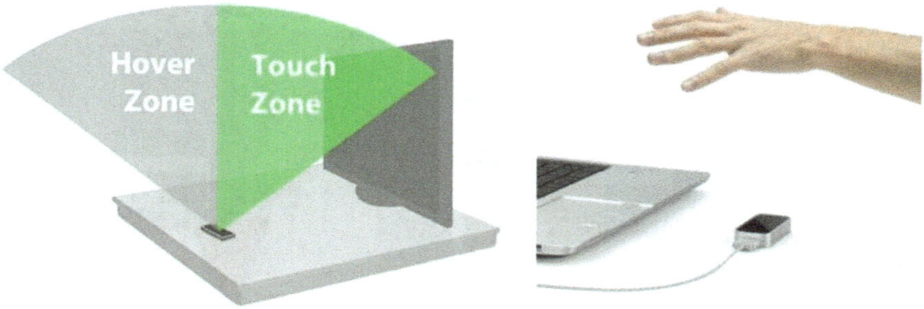

Fig. 15 Interaction area of leap motion (left), hand positions in interaction area (right)

has been conducted to educate and show children with ASD the proceedings of the serious game. The symbolism of the GameObjects has been briefly introduced to children. Thereafter, the serious game is played by them. Owing to certain physical disability and mental coordination experienced by the gamers, tapping motion will only be required to trigger the next move.

The position of the Leap Motion is adjusted accordingly to the range of motion of gamers. There are two areas zones for interaction—the Hover Zone and the Touch Zone shown in Fig. 15. The Hover Zone is used for aiming whereas the Touch Zone is used for creating touch events. The sensor can be adjusted during the trial session to detect the comfortable range of the hands before the actual game starts. To activate a collider, the gamers have to extend their hands towards the screen to enter into the touch zone, as shown in Fig. 15.

6.1 Scene Selection Menu

The selection menu has been designed with all buttons being evenly spaced out in the corresponding game scenes. The collider of the hand with the box will trigger the respective game scene to play. The background of the game scene selection menu is chosen to fit to the target audience, aiming to create a friendly learning environment.

Gamers can select the game scene and the life skills they want to learn. The movements of the hand will be reflected when the Leap Motion sensor prefabs captures the position and orientation of the fingers, palm, and wrists of the gamer.

6.2 Game Scene 1—Waking up in the Morning

The first game scene starts off with an alarm sound ringing to prompt the player to turn off the alarm. Gamers are required to tap the virtual button on top of the alarm clock, in order to turn off the alarm clock. The placements and sizes of the

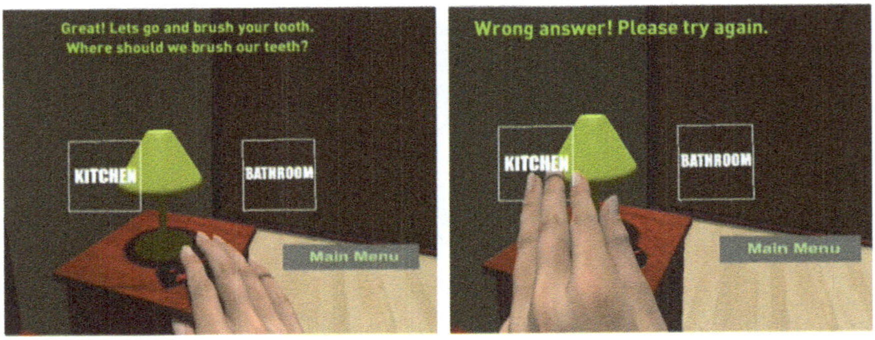

Fig. 16 Switch off alarm clock (left) and wrong selection of answer (right)

GameObjects are rendered and drawn to scale shown in Fig. 16. Gamers' position in the game scene is at the bedside when they just awaken.

Gamers need also to choose the right answer shown on the screen. Once all selections have been made, main menu button can then be selected by box collider, triggering to return to the main menu for selection of other game scenes.

The GUI texts are placed in the top-middle section of the game window. The colour contrast is chosen to be large to attract the attention of gamers. Short and concise messages are designed to simplify the learning process.

6.3 Game Scene 2—Brushing of Teeth

In the game scene 2, a step-by-step process is adopted in the design. The camera view covers the entire basin area of the virtual bathroom. This is to ensure that all GameObjects will be within reach of gamers in the game.

Gamers will be required to tap the GameObjects in a sequencing order. The step-by-step selection process is in the sequence: (1) Toothbrush, (2) Toothpaste, (3) Tap, (4) Cup. The visual feedback for the selection of GameObjects will be displayed on the screen. When a correct item is tapped, the name of the item will appear on the top right corner of the screen to indicate that it is successfully picked.

When all the items are selected following the correct sequence, a UI text will appear to prompt gamers to return to the main menu. A sound clip that mimics the sound produced when brushing of teeth will be played next. This indicates that brushing of teeth can take place after the right steps. A detected collision with the main menu button will redirect gamers back to the main menu to select other game scenes.

6.4 Game Scene 3—Making of Milo Drink

The camera view for making Milo drink features the kitchen in the virtual home. It displays the hot-water dispenser, Milo powder, spoon and cup used in the game scene. The background includes a tray with cake on it. This helps to set the environment for the game scene so that gamers will intuitively know that making of Milo drink is usually done in the kitchen.

Gamers are required to tap the GameObjects in a sequencing order. The step-by-step selection process is in the sequence: (1) Cup, (2) Milo Powder, (3) Hot-water Dispenser, (4) Spoon. The visual feedback for the selection of GameObjects will be shown on the screen. When a correct virtual item is tapped, the name of the item will appear on the top right corner of the screen to indicate that it is successfully picked. Upon completion of the task, an audial feedback of cup stirring will be played to indicate the success of task.

7 Experiment and Evaluations

7.1 Experiment Setup and Procedure

The participating school for these experiments is a special needs education school in Singapore, METTA School. The participants engaged in these experiments are with consent from their respective teachers and principals of METTA School. They are divided into two groups, namely the Game Group and the Control Group. While participants in the Game Group learn the skills from the developed Virtual Home serious game, participants in the Control Group learn the skills by conventional teaching.

The experiment of the Game Group is carried out in three stages as follows:

- *Stage* 1: *Demonstration*
  Participants watch a demonstration session to show them what should be done in the serious game. A story is given to narrate the scenario of home living situation. Questions regarding the participants' daily routine not only help to gather their participation but also provide an idea of what they can expect from this serious game. They are instructed to touch various objects shown in the game scene to test their understanding of the objects' symbolic meaning. Their ability to control the character and their ability to infer the purpose of the object are observed.
- *Stage* 2: *Game*
  Participants in the Game Group play the game, which takes them through the executable steps of their daily activities comprising *waking up in the morning*, *brushing of teeth* and *making of Milo drink*. Due to physical disability and restrictions or inability of comprehension, they can receive assistances whenever

required. Participants are observed during the play of the serious game. Their responses are assessed and evaluated by the staff of METTA School.

- *Stage 3*: *Assessment*
  The school staffs comprise of the teachers and Principal of METTA School. They are given an evaluation form to assess the participants' ability, level of engagement and the suitability of the game design. Participants in the Game Group are given a feedback form which measures their level of interest and satisfaction with the serious gameplay. Moreover, they are asked to indicate the effectiveness of the 'step-by-step' learning process in the given tasks.

The experiment setup for the Control Group is different from that of the Game Group.

The experiment of the Control Group of METTA School is carried out in two stages as follows:

- *Stage* 1: *Conventional Teaching*
  Participants are taught the three activities—*waking up in the morning*, *brushing of teeth* and *making of Milo drink*, using pictures and words as visual aids. The step-by-step process is taught to the participants verbally. They are taught once for all three activities. The pictures as visual aids displayed to them are shown in Figs. 17, 18 and 19.
- *Stage* 2: *Assessment*
  Participant is required to fill up an assessment form and write down the number of the items one by one in the correct sequence learnt in Stage 1. The answers are assessed to gauge their learning absorbability and capability to information taught in a conventional teaching method.

Upon completion of the experiment, participants are required to fill up an assessment form shown in Fig. 20.

Fig. 17 Visual aid for conventional teaching—waking up in the morning

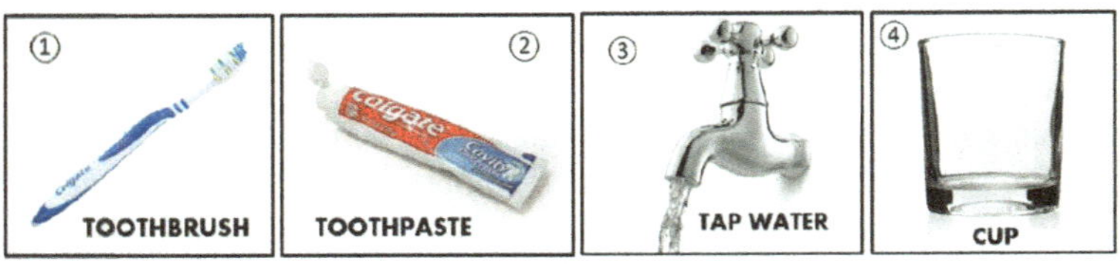

Fig. 18 Visual aids for conventional teaching—brushing of teeth

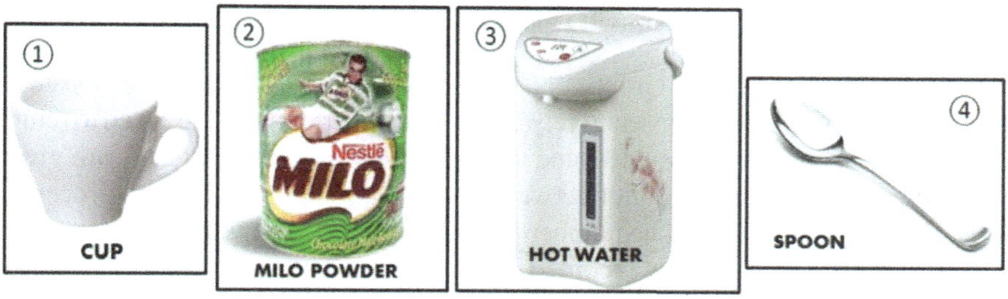

Fig. 19 Visual aids for conventional teaching—making of Milo drink

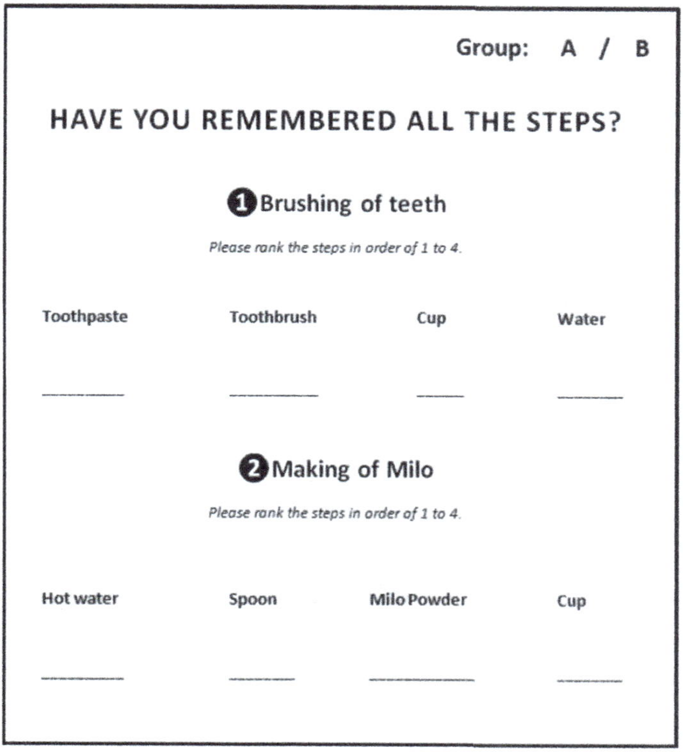

Group: A / B

HAVE YOU REMEMBERED ALL THE STEPS?

❶ Brushing of teeth

Please rank the steps in order of 1 to 4.

Toothpaste ______ Toothbrush ______ Cup ______ Water ______

❷ Making of Milo

Please rank the steps in order of 1 to 4.

Hot water ______ Spoon ______ Milo Powder ______ Cup ______

Fig. 20 Assessment form for the Control group and Game group

7.2 Experiment Execution

For the Game Group of METTA School, the experiment execution is conducted in the following settings.

- Location: METTA School IT Room.
- Participants: 6 students (4 males and 2 females), high-functioning ASD.

- Equipment: Laptop with Unity3D game engine, Leap Motion sensor.

Participants of the Game Group, under supervision of the teachers, are asked to play the three game scenes consecutively. After which, participants are asked to return to the main menu of the Virtual Home serious game. Assessment forms are given to them to assess their learning effectiveness of using VR's step-by-step processes of daily life skills teaching. The number of trials before attaining the correct answer for the step-by-step process is recorded and tabulated. The feedback for the Game Group in METTA School is conducted by a group of teachers. The teachers give their assessment and general comments on participants' performance and level of engagement during the experiment.

For the Control Group of METTA School, the experiment execution is conducted in the following settings.

- Location: METTA School classroom.
- Participants: 4 students (2 males and 2 females), moderate ASD.
- Equipment: Visual Aids and Assessment forms.

Two teachers are around for the teaching and assessment session. In a classroom setting, participants of the Control Group are taught the three game activities with the help of flashing, the visual aids, to them. After which, they will be required to fill in the assessment form. Teachers are around to guide them in answering the questions. Owing to their short attention span, they are not very receptive of the conventional teaching method. However, the experiment is still carried out as planned. Due to the privacy of participants, faces and particulars are not shown in Fig. 21.

Fig. 21 Control group—conventional teaching in classroom setting

7.3 Results and Analysis

The experiments are conducted at METTA School successfully. The feedback from the participants is that they enjoy the game. The experiment results and analysis are described next.

Participants in the Game Group are children with high-functioning autism which is at one end of the ASD spectrum. Their symptoms are less severe and have relatively fine motor skills. Hence, the general feedback is that the serious game may be too simple for these children. With practices, the participants may not be as engaged as when it is first introduced. This inevitably leads to the effectiveness of their learning process as it is more regurgitative.

The participants are very pleased with having to control the game with their hands. Their excitement has influenced their peers' enthusiasm towards the serious game. The participants in the Game Group can relate to the visual feedbacks and audial prompts. They are able to read and comprehend the meaning of the game objects. They can comprehend the rationale of the game and make themselves comfortable with the game by playing with the game scenes without much guidance. This has made it possible for the participants to be self-reliant while learning in a safe environment, when the serious games are designed to be engaging yet purposeful. The number of trials before attaining the correct answers is recorded in Tables 1, 2 and 3. This is to assess their first-hand understanding of the game design. When the number of trials is recorded as 1, it indicates that the participant gets it correct on the first attempt.

Observed from the results tabulated in Tables 1, 2 and 3, the average number of trials per step for all game scenes is below 3, with toothbrush having the highest number of trials of 2.67. This suggests that participants in Game Group are able to infer the meaning of the game objects by its placement, design and form. The modellings of the game objects have aided in their interpretation and correlation of

Table 1 Assessment of Game Group—waking up in the morning scene

Participant No	No. of trials	Observation and comments
Student 1	1	Accidental touch
Student 2	1	Correct interpretation
Student 3	2	Correct interpretation
Student 4	2	Correct interpretation
Student 5	1	Accidental touch
Student 6	1	Correct interpretation
Average number of trials	1.33	

Table 2 Assessment of game group—brushing of teeth scene

	Step 1—Toothbrush		Step 2—Toothpaste		Step 3—Tap		Step 4—Cup	
Participant No	No. of trials	Comment	No. of trials	Comment	No. of trials	Comment	No. of trials	Comment
Student 1	3	Insensitivity of game engine	1	Correct interpretation	1	Close proximity	3	Unable to detect hands
Student 2	2	Correct interpretation	1	Accidental touch	1	Correct interpretation	1	Correct interpretation
Student 3	1	Correct interpretation	1	Correct interpretation	2	Correct interpretation	4	Insensitivity of game engine
Student 4	1	Correct interpretation	4	Insensitivity of game engine	1	Correct interpretation	1	Correct interpretation
Student 5	6	Insensitivity of game engine	1	Correct interpretation	1	Correct interpretation	2	Correct interpretation
Student 6	3	Unable to detect hands	1	Correct interpretation	1	Correct interpretation	1	Correct interpretation
Average	2.67		1.5		1.17		2.0	

Table 3 Assessment of game group—Making of Milo drink scene

	Step 1—Cup		Step 2—Milo Powder		Step 3—Hot water dispenser		Step 4—Spoon	
Participant No	No. of trials	Comment	No. of trials	Comment	No. of trials	Comment	No. of trials	Comment
Student 1	1	Wrong interpretation	1	Accidental touch	1	Wrong interpretation	1	Accidental touch
Student 2	4	Unable to detect hands as fingers are closed	1	Wrong interpretation	1	Wrong interpretation	1	Wrong interpretation
Student 3	1	Wrong interpretation	1	Wrong interpretation	1	Wrong interpretation	3	Unable to detect hands as fingers are closed
Student 4	1	Wrong interpretation	4	Insensitivity of game engine	1	Wrong interpretation	1	Accidental touch
Student 5	1	Wrong interpretation	4	Insensitivity of game engine	1	Wrong interpretation	1	Wrong interpretation
Student 6	2	Mistaken Milo powder for cup	1	Wrong interpretation	1	Wrong interpretation	1	Accidental touch
Average	1.67		2.0		1.0		1.33	

the objects' naming and placement. However, there are some exceptions like the can of Milo powder is mistaken to be a cup by some participants. Hence, more effects in graphic rendering for the texture should be done.

There are numerous feedbacks on the inability to detect participants' hands due to the insensitivity of the Leap Motion sensor. Moreover, this can be attributed by the distance, speed, and the orientation of the hand. This is one limitation that is apparent across all game scenes. There are trials that are passed due to the close proximity of the objects which may not be reflective of the participants' understanding. Hence, the placement of the game objects should be spaced further apart. The Box Collider could be enlarged to enhance its sensitivity to the participants' hand movements.

Moreover, observed from the assessment forms given to the participants, they have shown a significantly better performance in terms of remembering the steps for the tasks. Figure 22 shows the graphical results of the number of correct answers given by the participants who played the game.

The participants in the Control Group are taught the routine life of students in terms of waking up in the morning, brushing of teeth and making of Milo drink. The teaching is conducted in a conventional classroom setting, with the help of visual aids. Upon completion, the participants are asked to complete the assessment form.

Each participant has four answers for each game scene, and they are required to rank the steps of the tasks accordingly. The number of correct answers is assessed. The graphical results are shown in Fig. 23. The results of the assessment show that only 7 out of 32 possible answers are correctly filled in by the participants. Hence, only 21.88% of the answers are correct.

This may be because the participants are unable to remember the steps as taught earlier on. Moreover, owing to their short attention span, they are not able to devote their full attention to the teachers' words. They tend to get easily distracted.

Hence, conventional classroom teaching using images and flashcards may not be efficient for children with ASD to learn certain life skills.

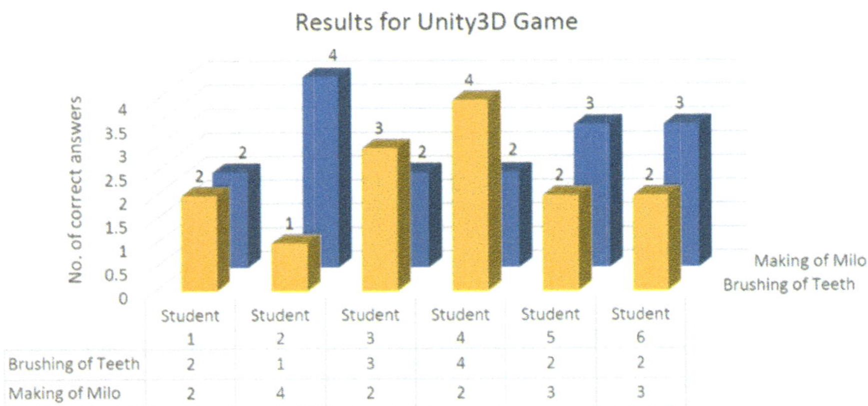

Fig. 22 Game Group—number of correct answers for each step in two tasks

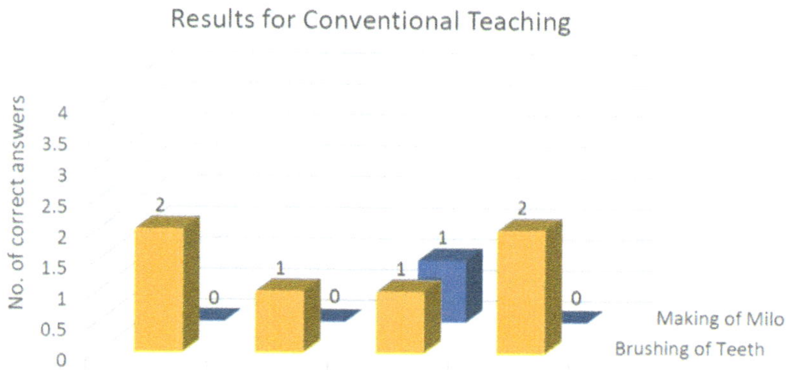

	Student 1	Student 2	Student 3	Student 4
Brushing of Teeth	2	1	1	2
Making of Milo	0	0	1	0

Fig. 23 Control group—number of correct answers for each step in two tasks

It is observed that the participants in the Game Group are able to perform much better than those in the Control Group who undergo conventional teaching with visual aids. There is a significant increment in the number of correct answers recorded. 62.5% of answers are correct for the participants in the Game Group undergoing the Unity3D serious game, whereas conventional teaching method has only resulted in 21.88% of correct answers recorded. This has shown that participants are able to remember the steps better when they are more engaged. They have hands-on activity to capture their attention for a longer period of time.

There are varying number of trials before attaining the correct answers. This is due to various reasons like technical or software insensitive, unsuitable placement of game objects and lack of comprehension from them. However, with more practice and slight modifications to the current game design, they will benefit more from this serious game.

Observed from the serious game, children with ASD are very participative and engaged during the demonstration and game-playing stages. They clap and cheer for their friends when the right answer is chosen. They can relate to the feelings of the visual feedback provided on the 'tick' and 'cross' that represents right and wrong answers, respectively. The most encouraging result observed is the participants' amazement at seeing the hand on the screen and their active participation during the serious game. Most importantly, they understand the objective of the serious game well.

Generally, the teachers proposed for more tasks to be developed with the use of the Leap Motion sensor. Owing to the participants' high functionality in the ASD spectrum, suggestions for more discrete hands gestures to be incorporated in the game design are received. The feedback and comments given by the staff of METTA School can be found in Table 4.

Table 4 Feedback and comment given by staffs of METTA School

No	Questions	Consolidated staffs comments
1	Students' level of engagement (1—Least Interested, 5—Most Interested)	4
2	Do you think the children understood how to control the hand?	Yes
3	Do you think the children will be able to learn from the step-by-step process?	Yes, definitely with more practice
4	Do you think the children was able to follow the sequence?	Yes
5	Do you think children learnt from the game? (1—Strongly Disagree, 5—Strongly Agree)	3
6	Do you think the children understood the meaning of the game?	Yes
7	Do you think the checklist was useful?	Yes
8	Do you think the words are useful in guiding the children?	Yes
9	Do you think this interactive session was beneficial for the children?	Yes
10	Comments on children's responses	– Students were excited when they see the hand – Some assumed that the screen was a touch screen – Students want more of such games for different skills
11	Enhancement feedback	– Animation to show the process—Zoom in and Zoom out effect – Discrete hand gestures to be captured (e.g. squeezing of toothpaste etc.)

Some feedback is given by the teachers to further improve the game design, in order to better cater to their students' needs. The checklist in the serious game may not be useful to the participants as it is not obvious or emphasized, although the GUI is useful in guiding them in the learning process. The teachers also have suggested for a more immersive experience to be provided.

According to the suggestions, the display of the words could be larger, and items could be spaced wider apart for participants who are physically impaired to reach. Moreover, voice-over instructions will be ideal for them to feel more immersive during the serious game. Additionally, more sound effects could be inserted to enhance the audial feedback and to provide a more conducive and immersive learning experiment for the participants.

As this work serves only as a prototype serious game, further implementations need to be done to enhance the children's learning capacity and drive. An incorporation of more audial instructions and accompanying background music will help create a more inclusive learning environment. Symbols and buttons used could be emphasized by increasing the size and adjusting of the colours. Emphasis like entrance and exit animation could be added to attract the attention of children with ASD.

Besides these tasks featured in this work, further implementations could look into a daily routine life of students, which includes the preparation of meals or preparation for school. These activities are essential in their life and hence, allowing learning through serious games would be a safer and more effective solutions. Animation of the tasks could be done to illustrate the participants' expected actions when the activity is accomplished.

The aforementioned improvements could have been executed with permissible time and deeper understanding of the Unity3D game engine and the Leap Motion API.

8 Conclusions

8.1 Contributions

The purpose of this research is to provide children with ASD an alternative learning platform that is engaging, captivating and meaningful to them. In light of this, the serious game may be used to enhance their learning capability and depth of learning by broadening their scope. More delicate and dangerous tasks may now be learnt through the implementation of VR environment.

Parents and educators often have concerns about teaching children with ASD how to respond appropriately when executing certain tasks. For accident-prone tasks, practice can be problematic. With the help of VR, it addresses common learning problems for children with ASD. These benefits include repeatable, consistent practice with predictable responses and clear on-screen guidance.

Safer learning environment, controlled input stimuli and complexity helping with appropriate task focus, and strong engagement value are effective teaching formats in the virtual scenes. The experiment results indicate that VR is a highly applicable and effective platform for children diagnosed with ASD to learn life skills. It is conclusive that game designs with great emphasis on the appearance of the objects coupled with good audial and visual feedbacks will enhance the efficacy of their learning ability. It is also shown that the control of the hands is intuitive when the player witnesses the reciprocated mimicked actions on screen.

8.2 *Limitations and Future Improvements*

Though this research is very promising, there are some limitations and drawbacks to the game design, development, and implementation. One major drawback is the lack of accuracy of the Leap Motion sensor. The loosing track of fingers by the Leap Motion sensor is evident during the experiment in METTA School. This is not beneficial as children with ASD often get distracted with their chain of thoughts disrupted. This has impeded their learning process as the flowlines of the serious game is not optimized. However, despite the drawback of the Leap Motion sensor, it can track different hand gestures. Since the Leap Motion API can detect precise finger movements, a plausible development of VR games for children with similar disabilities would be to encourage children to execute precise gestures and actions like squeezing, grabbing, and turning of knobs.

In addition to the game-development improvements, there are other improvements for consideration. The design of the virtual home is done in Autodesk Maya which allows for graphic designing without precise dimensions. This has made the designing process much more efficient. However, owing to time constraints and lack of expertise, minimal furniture and household appliances are designed. In order to bring more realism to the serious games, more household appliances and decorations could be designed and rendered.

Owing to the importation of files with extensions of.mb or.ma from Autodesk Maya into the Unity3D game engine environment, certain textures and shades of the game objects are not recognized by Unity3D game engine platform. Hence, this had resulted in a lack of resemblance of some game objects. For example, a case in point is the Milo container. Hence, this lack of compatibility has lessened the realism and effectiveness of the serious game.

In addition, the number of children with ASD participate to the experiments are limited due to the clashes of their curriculum and the availability of their schedules. Hence, the results and analysis may not be an accurate reflection of its effectiveness as the sample size is small.

Acknowledgements The authors would like to thank the students, teachers, staffs, principal, and parents of METTA School for their support, help, and feedback in this research work.

References

1. Bartoli, L., Corradi, C., Garzotto, F., Valoriani, M.: Exploring motion-based touchless games for autistic children's learning. In: 12th International Conference on Interaction Design and Children (2013). https://doi.org/10.1145/2485760.2485774.
2. American Psychiatric Association.: Diagnostic and statistical manual of mental disorders, 5th edn. (2013). https://doi.org/10.1176/appi.books.9780890425596.
3. Weston, C.: Four social brain regions, their dysfunctions, and sequelae, extensively explain autism spectrum disorder symptomatology. Brain Sci. **9**(6), 130 (2019). https://doi.org/10.3390/brainsci9060130

4. Landa, L.: Early communication development and intervention for children with autism. Mental Retard. Dev. Disabilit. Res. Rev. **13**(1), 16–25 (2007). https://doi.org/10.1002/mrdd.20134
5. Cai, Y., Chiew, R., Nay, Z.T., Indhumathi, C., Huang, L.: Design and development of VR learning environments for children with ASD. Inter. Learn. Environ. **25**(8), 1098–1109 (2017). https://doi.org/10.1080/10494820.2017.1282877
6. Zhu, L., Cao, Q., Cai, Y.: Development of augmented reality serious games with a vibrotactile feedback jacket. Virtual Reality & Intelligent Hardware. **2**(5), 454–470 (2020). https://doi.org/10.1016/j.vrih.2020.05.005
7. Didehbani, N., Allen, T., Kandalaft, M., Krawczyk, D., Sandra, C.: Virtual reality social cognition training for children with high functioning autism. Comput. Human Behav. **62**(C), 703–711 (2016). https://doi.org/10.1016/j.chb.2016.04.033
8. Kandalaft, M.R., Didehbani, N., Krawczyk, D., Allen, T., Sandra, C.: Chapman virtual reality social cognition training for young adults with high-functioning autism. J. Autism Dev. Disord. **43**(1), 34–44 (2013). https://doi.org/10.1007/s10803-012-1544-6
9. Maskey, M., Lowry, J., Rodgers, J., McConachie, H., Parr, J.R.: Reducing specific phobia/fear in young people with Autism Spectrum Disorders (ASDs) through a virtual reality environment intervention. PLoS One **9**(7) (2014). https://doi.org/10.1371/journal.pone.0100374o.
10. Henry, M.: Senior project report: MIDI flapper, a leap motion MIDI controller. California Polytechnic State University, San Luis Obispo (2014)
11. Potter, L.E., Araullo, J., Carter, L.: The Leap Motion controller: a view on sign language. In: 25th Australian Computer Human Interaction Conference: Augmentation, Application, Innovation, Collaboration (2013). https://doi.org/10.1145/2541016.2541072.

Learning to Cross Roads Through VR Playing

Qingqing Zhang, Qi Cao, and Yiyu Cai

Abstract Research shows that virtual reality (VR) can be used as a teaching aid for children with autism spectrum disorder (ASD). In special needs education, one of the difficult parts is to increase the time of concentration of children with ASD in their learning. Using VR technology with 3D visualization and realistic rendering, the learning concentration as well as the teaching efficiency can be improved in special needs education. VR technology can also be extended to other applications of special needs education. This chapter presents the research work help children with ASD to learn basic skills of road crossing in traffic junctions with real-world scenarios simulated. The design and development of a road crossing serious game are described for independent learning of children with mild and moderate ASD. For children with severe ASD, teachers may need to give them additional aids and instructions in the learning process when they are using the road crossing serious game. Experiment results are promising. Currently, this serious game is designed for the left-hand traffic system like Singapore and UK. Right-hand traffic system will be developed in future.

Keywords Independent learning · Self-care skills · Basic life skills · Road crossing · Traffic rules

Q. Zhang · Y. Cai (✉)
School of Mechanical and Aerospace Engineering, Nanyang Technological University, Singapore 639798, Singapore
e-mail: myycai@ntu.edu.sg

Q. Cao
School of Computing Science, University of Glasgow, Singapore Campus, Singapore 567739, Singapore
e-mail: qi.cao@glasgow.ac.uk

Y. Cai and Q. Cao (eds.), *When VR Serious Games Meet Special Needs Education*, Gaming Media and Social Effects, https://doi.org/10.1007/978-981-33-6942-9_4

1 Introduction

1.1 Background

Autism spectrum disorders (ASD) is a neural development disorder. Normally, children with ASD show much lower performance in both physique and intelligence. Children with ASD have impairments in communication and social interaction. They often have sensory processing difficulties and attention abnormalities predisposing them to process information differently and respond in unusual ways. Sadly, there is currently no known cure. Research shows that children with ASD still can learn and handle skills after numerous repeating and practicing.

Children younger than 10 years of age and adults aged above 65 are groups with the highest percentage of pedestrian fatalities in traffic accidents [2]. About 60% of pedestrian injuries occur when children cross a road at road junctions [8]. Particularly, children with ASD face a high risk of being injured when crossing a street [3]. Parents have indicated street-crossing skills as an area of concern for their children with ASD [9]. Road safety education is an essential skills component of children's education, especially those with ASD to handle traffic situations safely [6, 9]. Methods for teaching street-crossing skills conducted in classroom environment have been shown to be ineffective to cross a real street [3]. There is a need for better teaching methods for the road safety and skills, using innovative technologies such as virtual reality (VR) [9].

It is challenging to teach and help children with ASD learning the basic life skills and self-care skills. Traffic crossing is a common skill to most of us in daily life. Understanding the traffic signals and rules is the basic life skill to live in a metropolis like Singapore. It is dangerous for children with ASD to learn street-crossing skills directly in natural environment [1, 3]. Normal people acquire this simple skill in primary school or even kindergarten. However, the learning of street-crossing knowledge and skills in a safe manner are difficult for children with ASD [3].

Research shows that serious games with VR and user interactivity can improve the learning efficiency of children with ASD. Considering the training on real streets environment poses safety concerns, VR may be a better training medium for this road crossing skills [1].

A VR environment is introduced to train 7–8 years old children to gain street-crossing knowledge [8]. The experiment outcomes show children trained in the virtual environment yields consistently safer pedestrian behavior than those trained in conventional methods. A desktop PC version serious game is developed by Josman et al. [3] to train children with ASD street-crossing skills. A VR serious game to train pedestrian skills including street crossing and street navigation is depicted by Saiano et al. [7]. A PC-based fire and street safety VR game for children with ASD is introduced by Strickland et al. [11]. A VR game is developed to teach individuals with ASD how to take buses and use public transportation systems [10]. Miller et al.

[5] present a VR game using a smartphone and Google Cardboard device, to teach children with ASD basic air travel skills from check in to boarding an airplane.

There are many games developed using Microsoft Kinect motion sensor with gesture inputs [4]. In this serious game, a touch screen and a Microsoft Kinect device are used. Unity3D game engine, a cross-platform game engine with built-in integrated development environment (IDE), is applied to develop this serious game. It has programming libraries and developing tools in addition to JavaScript, C#, and Boo for scripting. JavaScript is chosen for this serious game development with keyboard inputs as a basic interface.

As there is no direct adaptor between Unity3D and Microsoft Kinect, the Flexible Action and Articulation Skeleton Toolkit (FAAST) is used acting as the third-party plugin [12]. It is an open source software for the interface between keyboard inputs and Kinect interaction. The FAAST tool is very useful and can be customized for different purposes.

The difficult parts of the animation development are the character riggings and body movements. Although Unity3D has its own animation tool, but undoubtedly, Autodesk Maya is more widely used, because of two innovative tools. One is the skeleton and skin binding tool, and the other is weight paint tool. All the animation in this serious game is developed on the Autodesk Maya platform.

1.2 Scope and Objectives

The scope of this work is to create a VR serious game to simulate traffic scenarios at a virtual traffic junction. The serious game supports two interactive capabilities using Microsoft Kinect device and the touch screen. A traffic junction and a zebra scenario are simulated in an interactive serious game for children with ASD to acquire the basic road crossing skills. Actions such as button pressing, traffic light waiting, and gesture signals can be made by the children when learning road crossing.

1.3 Organization of the Chapter

The remaining parts of the chapter are organized as follows. Section 2 presents the development details of the road crossing serious game. Section 3 describes the game illustrations. The future development and conclusions are depicted in Sect. 4.

2 Game Design and Development

2.1 Concept of the Road Crossing

Singapore has a left-hand traffic system following the British traffic rules. There are some differences to the mainstream right-hand traffic system. For example, when pedestrians prepare to cross the road in the left-hand traffic system countries, they should look to the right first, then left, before they start crossing roads. While in those right-hand traffic system countries, pedestrians should look to the left first, then right, before they start crossing roads. Some people may ignore the differences. But for a better training purpose, this serious game strictly follows the international standard of pedestrians traffic rules. It aims to ensure the highest safety for children with ASD in learning and training. The basic rules of traffic light signal are as follows.

(1) *Green walking man/Green Light:* Cross with caution. Pedestrians have the highest priority on the zebra.
(2) *Red standing man:* Do not cross.
(3) *Flashing green walking man:* Continue crossing if already in the intersection; otherwise do not start to cross.

Some pedestrians may forget the third point. Some traffic accidents occurred at the traffic junction are caused by the rushing behavior when the green light is flashing.

2.2 Modeling of the Traffic Junction Scene

The 3D models of this serious game are all created in the Autodesk Maya and imported into Unity3D in.*mb* file format. The importing and exporting processes are very convenient between Unity3D and Autodesk Maya. Some examples of the 3D models developed in this serious game including streetlight, vehicle, pedestrian, and buildings are shown in Fig. 1.

3D designs using software such as 3ds MAX or Solidworks are exportable to Unity3D. The.*fbx* file is a popular format for the importing, especially for those 3D models with animations. The 3D models designed in Autodesk Maya can be exported into Unity3D directly. But there are still some problems in the importing process.

The smooth effect in the Autodesk Maya files cannot be imported into the Unity3D. The only way to add the smooth rendering in Unity3D is to increase the polygon number, which increases the files size and affects the running speed of the Unity game.

The frame frequency is different between Autodesk Maya and Unity3D. It means the animation developed in Autodesk Maya will be slower in the Unity3D platform. The exact ratio of the frame frequency differences is not constant, as the updating frequency of the Unity3D is not fixed. For this problem, the FixedUpdate() function needs to be used.

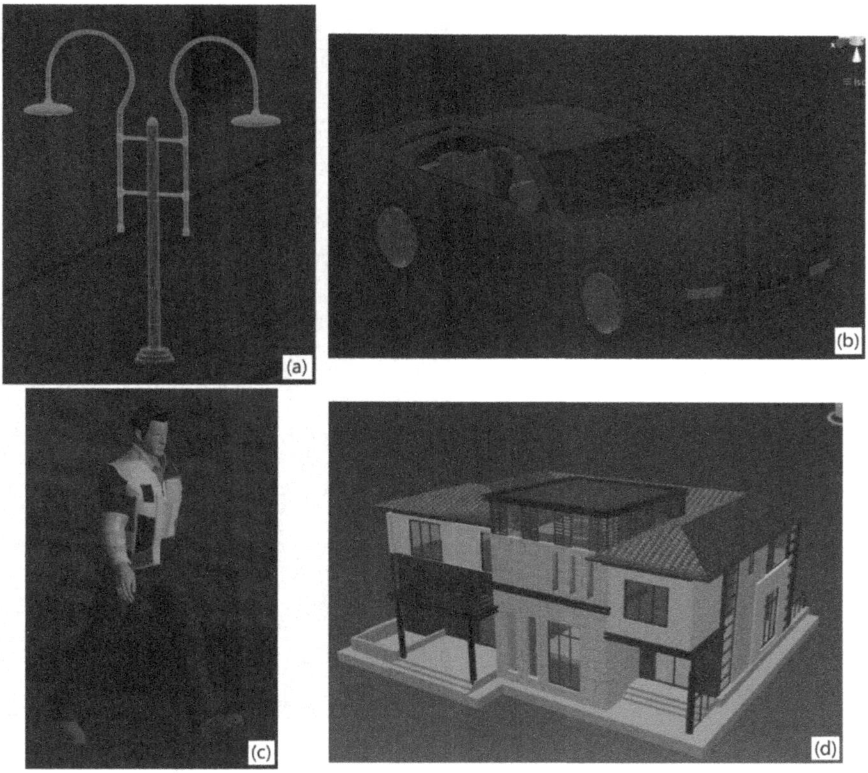

Fig. 1 3D models designed in this serious game, **a** streetlight, **b** vehicle, **c** pedestrian, **d** buildings

A small imperfection in the Autodesk Maya model file might become obvious after importing to the Unity3D. For example, the break of the surface or skin of the 3D models in Autodesk Maya will be amplified in the Unity3D. After the importing, there could be also some other information lost such as texture, rendering, and materials. Hence, it needs to enhance the surface or texture in the Unity3D, which can ensure the perfect surface matching to avoid similar problems.

2.3 Animation of Character Body Movement

The animation design of character body movements is a difficult yet time-consuming part of the game development. The standard of a good game is always highly related

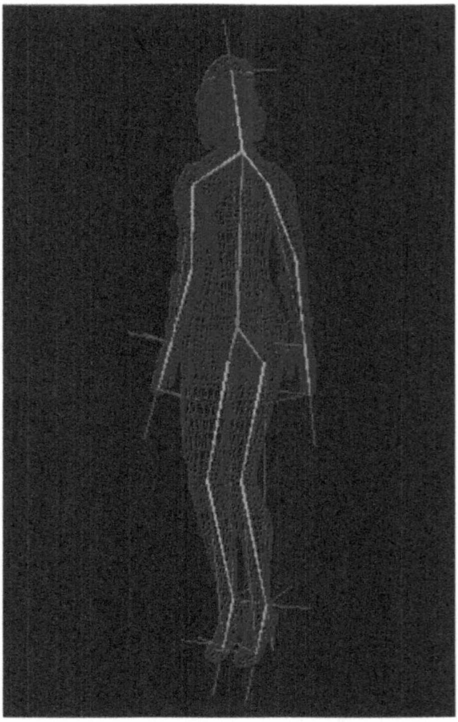

Fig. 2 Skeleton binding

to the animation and realistic visualization quality. There are two powerful tools in Autodesk Maya that can be used in the game design.

Skeleton and Skin binding

The skeleton binding in Autodesk Maya is shown in Fig. 2. As its name suggests, this tool is to create the skeleton of the 3D models. The skeleton will act as the controller of the body motion or the facial expression. In the development of this serious game, some design skills have been adopted as follows.

In the skeleton creation process, the skeleton joint has five degrees of bending, instead of straight. It can bring convenience in the later binding process.

The skeleton is fit to the outer surface as perfect as possible. Otherwise, it will cause the skin broken in the later rigging process.

Each time when the skeleton is created, a baking process is followed. It can ensure the smooth binding in the game design.

As mentioned in Sect. 2.2 previously, the frame frequency is slightly different between the Autodesk Maya and Unity3D. It might cause some jerking motion in the later animation process. To prevent this problem, the start key frame and end key frame of the animation should be identical.

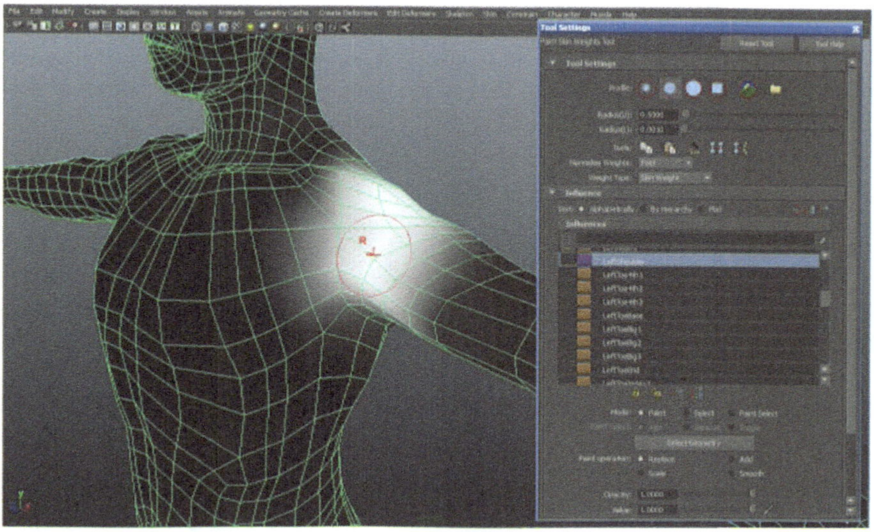

Fig. 3 Weight paint tool

In the design process, it is necessary to organize the skeleton in a hierarchy form, such as spine, arms, fingers. By this approach, the designers can find the suitable joints faster and improve the working efficiency.

Weight Paint Tool

Weight paint is the unique animation development tool in Autodesk Maya. The amount of influence to a joint can be customized on a given surface. The whiter the color, the greater the effect from a joint. An example of design interface is shown in Fig. 3.

Some design techniques have been used in the development of this serious game. Before the weight of a joint is painted in Autodesk Maya, the whole body is flooded with a zero-weight amount. It is necessary for this step, as sometimes designers may not notice some unexpected small white spots painted accidently on some unrelated body part. It is difficult to paint an inner surface, which is inconvenient to access from outside. For example, it is not easy to paint the oral cavity in order to animate the mouth movement. In this work, the jaw joint is moved a bit lower. Using this method, the oral cavity can open simultaneously when the mouth area is painted.

2.4 Programming in Unity3D

Unity3D is a cross-platform game engine with built-in IDE. There are many run-time classes under the namespace "UNITYENGINE". It has a very powerful 3D physical engine, which can simulate the dynamic motion realistically such as collision, dropping, and explosion. All those run-time classes are organized in hierarchy. A Unity3D script acts as a component of a certain game object. A single script can be attached to multiple game objects. This method can help to save a lot of development time, with enhanced working efficiency.

2.5 FAAST Mapping

Another part of the design of this serious game is to combine the Unity3D game engine to the Microsoft Kinect, which are linked by a third-party plugin FAAST. Figure 4 shows the interface of the FAAST software. In FAAST, the gesture inputs can be customized. After the presetting of FAAST, gesture inputs can then be changed

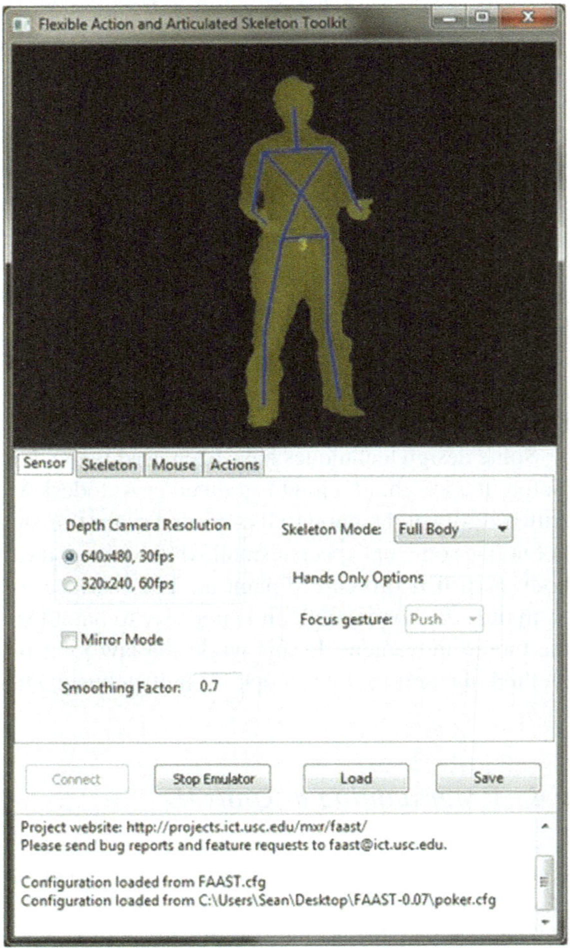

Fig. 4 FAAST interface

into keyboard inputs. By this approach, the Unity3D game engine can be finally combined with Microsoft Kinect.

3 Game Demonstration

Once the development of the road crossing serious game is done, children from Asian Women's Welfare Association (AWWA) School, a special needs school in Singapore, are invited to the game trial. Children with ASD from AWWA School become gamers. Their teachers are involved in the video recording during the video demonstration stage. The teachers of AWWA School also provide instructions and help children with ASD in the game stage.

Gamers can choose different resolutions and screen ratios for the best fit display, before starting the road crossing serious game. Once launching the serious game, gamers can choose different options in the main menu as shown in Fig. 5.

The learning process for road crossing serious game activities has been conducted in three stages as follows.

(1) *Stage 1: Game story* - Children with ASD are introduced the in-game character and the background of the game story. The in-game character wants to take a bus to the school, while the bus stop is on the opposite side of the road. Children need to press the traffic light button, wait for the green man, and then cross the road junction. The game story can help children to link the serious game with real-world scenarios.
(2) *Stage 2: Video demonstration* - A video tutorial is recorded by a teacher of AWWA School. It demonstrates to children what to do in different scenarios in

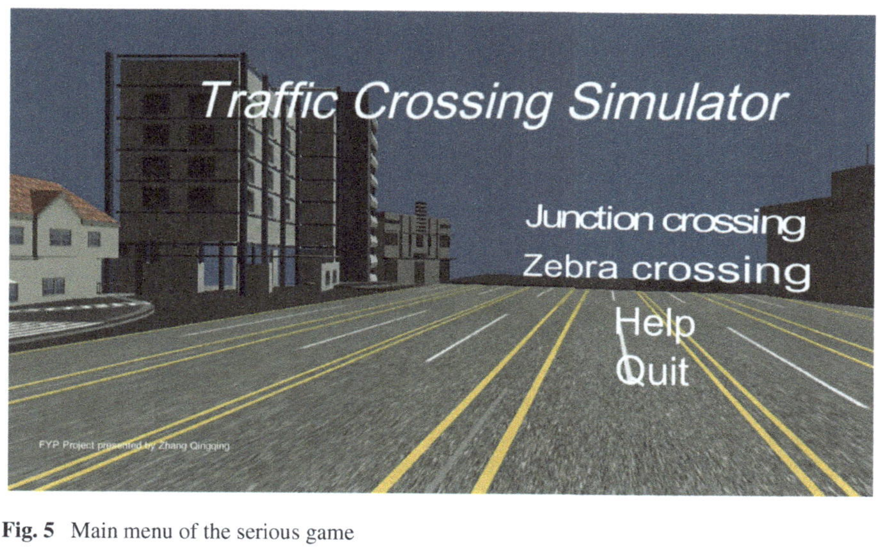

Fig. 5 Main menu of the serious game

Fig. 6 Red standing man scenario—do not cross

the serious game. Gamers can follow the instructions from the video tutorial and use the knowledge learnt in the game stage later.

(3) *Stage 3: Game* - Gamers start playing the road crossing serious game one by one. It takes them through the steps on how to cross a road at the traffic junction. At this stage, they can practise what they have learnt from the previous two stages. The game playing could reinforce their learning process and knowledge of what to do in the real-life environment.

According to the traffic rules described in Sect. 2.1, there are three scenarios in the road crossing serious game developed in this work, which are *green walking man, flashing green walking man,* and *red standing man.* These three scenarios are shown in Figs. 6, 7 and 8 separately.

In the serious game, when in-game character arrives at the traffic junction and the traffic light is a red man, gamers need to press the traffic light button to trigger the change of the traffic light. In Singapore traffic context, once the traffic light button is pressed by a pedestrian, the traffic light will change in certain time frame. It is to tell the traffic light system that some pedestrians are waiting to cross this road junction.

When the traffic light changes and the green walking man is on, gamers can now start crossing the road junction. To achieve it, they need to lean forward to trigger the walk forward motion in this serious game. The lean forward pose of gamers is set to control the walk forward command.

Observed from the experiment and the three stages learning process, children with ASD have shown their good interests when learning to cross roads. They enjoy playing the road cross serious game. According to the feedback of the children, the game stages help them to better understand the traffic rules for road crossing, which is usually difficult to memorize in traditional classroom education.

Fig. 7 Green walking man scenario—cross with caution

Fig. 8 Flashing green walking man scenario—continue crossing if already in the intersection; otherwise do not start to cross

4 Conclusions and Future Improvement

In this work, a VR serious game is developed, to teach children with ASD how to cross the road junctions safely. Although there are many rooms to improve with the developed serious game, children with ASD from AWWA School are excited in the learning process. Because of the confinement of the game course, they can pay good attention to learn the skill during the play based learning process. The responses and feedback from AWWA School are quite good. Most children with ASD can finish the serious gameplay by themselves without guidance after practicing for several times.

It is observed that some children with ASD from AWWA School can even complete the serious game successfully without the video tutorial at the video demonstration stage. This indicates some of them might be able to cross the traffic junction in the virtual environment by their own, which is desired for the learners to have this skill in the real life. The VR serious game could better help them in the learning journey, as it provides virtual environment to mimic the real-life scenarios.

There are some future improvements for the developed VR serious game in this work as follows.

(1) To achieve a better realistic environment in the serious game, the animation could be designed in more sophisticated and delicate ways.
(2) The current version of the VR serious game is only designed for left-hand traffic system, which will limit the learner population. Moving forward, the right-hand traffic system can be adopted in the game design. Learners can choose which traffic system from the main menu.
(3) The video tutorial in the demonstration stage needs to be improved both in video resolutions and sound quality. It could enhance the learning outcomes and better attract the attentions of children with ASD.

Acknowledgements The authors would like to thank the students, teachers, staffs, principal, and parents of AWWA School for their support, help, and feedback in this research work.

References

1. Dixon, D.R., Miyake, C.J., Nohelty, K., Novack, M.N., Granpeesheh, D.: Evaluation of an immersive virtual reality safety training used to teach pedestrian skills to children with autism spectrum disorder. Behav. Anal. Pract. (2019). https://doi.org/10.1007/s40617-019-00401-1
2. EU Commission.: Road safety: age groups most involved in fatal crashes. https://ec.europa.eu/transport/road_safety/specialist/knowledge/pedestrians/crash_characteristics_where_and_how/age_groups_most_involved_in_fatal_crashes_en (2020)
3. Josman, N., Ben-Chaim, H.M., Friedrich, S., Weiss, P.L.: Effectiveness of virtual reality for teaching street-crossing skills to children and adolescents with autism. Int. J. Disabil. Dev. Educ. **7**(1), 49–56 (2008). https://doi.org/10.1515/IJDHD.2008.7.1.49
4. Lun, R., Zhao, W.: A survey of applications and human motion recognition with Microsoft Kinect. Int. J. Pattern Recognit. Artif. Intell. **29**(5), 1–48 (2015). https://doi.org/10.1142/S0218001415550083
5. Miller, I.T., Wiederhold, B.K., Miller, C.S., Wiederhold, M.D.: Virtual reality air travel training with children on the autism spectrum: a preliminary report. Cyberpsychol. Behav. Soc. Network. **23**(1). https://doi.org/10.1089/cyber.2019.0093 (2019)
6. Muir, C., Devlin, A., Oxley, J., Kopinathan, C., Charlton, J., Koppel, S.: Parents as role models in road safety. Report of Accident Research Centre, Monash University (2010)
7. Saiano, M., Garbarino, E., Lumachi, S., Solari, S., Sanguineti, V.: Effect of interface type in the VR-based acquisition of pedestrian skills in persons with ASD. In 37th Annual International

Conference of the IEEE Engineering in Medicine and Biology Society (EMBC), Milan, Italy. https://doi.org/10.1109/embc.2015.7319693 (2015)

8. Schwebel, D.C., McClure, L.A., Severson, J.: Teaching children to cross streets safely: a randomized controlled trial. Health Psychol. **33**(7), 628–638 (2014). https://doi.org/10.1037/hea0000032
9. Self, T., Scudder, R.R., Weheba, G., Crumrine, D.: A virtual approach to teaching safety skills to children with autism spectrum disorder. Top. Lang. Disorder **27**, 242–253 (2007). https://doi.org/10.1097/01.TLD.0000285358.33545.79
10. Simões, M., Bernardes, M., Barros, F., Castelo-Branco, M.: Virtual travel training for autism spectrum disorder: proof-of-concept interventional study. JMIR Serious Games, **6**(1) (2018). https://doi.org/10.2196/games.8428
11. Strickland, D.C., McAllister, D., Coles, C.D., Osborne, S.: An evolution of virtual reality training designs for children with autism and fetal alcohol spectrum disorders. Top. Lang. Disord. **27**(3), 226–241 (2007). https://doi.org/10.1097/01.TLD.0000285357.95426.72
12. Suma, E.A., Lange, B., Rizzo, A., Krum, D.M., Bolas, M.: FAAST: the flexible action and articulated skeleton toolkit. IEEE Virtual Real. (2011). https://doi.org/10.1109/VR.2011.5759491

Virtual Pink Dolphins and Lagoon

Dongjun Lu, Qi Cao, and Yiyu Cai

Abstract Children including those with autism spectrum disorders (ASD) like to play with pets. Dolphin-assisted therapy (DST) as one type of pet-assisted therapy is already introduced to help children with ASD. However, DST is fairly costly. Given a large number of children with ASD interested in DST, it is difficult to have sufficient dolphins available to provide the service. Despite popular among children with ASD, DST is often criticized by animal rights organizations. Sometimes, it can also become challenging to use the captive dolphins for therapeutic purpose for those weak children with ASD. The aim of this research work is to develop an innovative solution for virtual pink dolphin-assisted therapy for children with ASD. As species endangered, pink dolphins are extremely well received by children. This chapter presents a serious game for children with ASD to play with virtual dolphins and improve their communication and learning. The focus of this work, however, will be on the object modeling and design of virtual pink dolphin serious game, including concept design, texturing, and animations with some cases analysis. The pink dolphin serious game is designed for single player.

Keywords Dolphin-assisted therapy · Pink dolphins · 3D VR games · Serious game design · Virtual lagoon modeling

---

D. Lu · Y. Cai (✉)
School of Mechanical and Aerospace Engineering, Nanyang Technological University, Singapore 639798, Singapore
e-mail: myycai@ntu.edu.sg

Q. Cao
School of Computing Science, University of Glasgow, Singapore Campus, Singapore 567739, Singapore
e-mail: qi.cao@glasgow.ac.uk

Y. Cai and Q. Cao (eds.), *When VR Serious Games Meet Special Needs Education*, Gaming Media and Social Effects, https://doi.org/10.1007/978-981-33-6942-9_5

1 Introduction

1.1 Background

Children with autism spectrum disorders (ASD) suffer impairments in communication and social interaction, with sensory processing difficulties and attention abnormalities, thus, they process information differently and respond in unusual ways [4, 5, 12]. Dolphin-assisted autism therapy, which encounters between dolphins and children with ASD have been explored in some research works [10]. Children with ASD may feel safe with the presence of pets, such as dolphins in therapy environments [8]. Dolphins are often considered as friendly aquatic mammals of big shape, intelligent, curious, non-threatening expression, cooperative and playful attitude, and accepting physical contact [15]. Dolphin-assisted therapy involves interaction between children with ASD and dolphins. Dolphins are willingly trainable for establishment of their relationships with humans, which has possible therapeutic effects for children [2]. Research works show animal-assisted therapy, such as dolphin-assisted therapy could help children with ASD to improve their social interactions and communications, especially with the aids of instructors [7, 8]. It also helps children with ASD to enhance the focus and social environment awareness [8, 13]. Therapists work alongside dolphins and their trainers are assisting children to achieve learning and therapeutic objectives [6].

Indo-Pacific humpback dolphins, also known as pink dolphin, are a species meeting the International Union for Conservation of Nature (IUCN) Red List requirements for Vulnerable under criteria A4cd [9, 11]. Pink dolphins, the friendly and social creatures, can be found off the coast of China, Singapore, Thailand, and Vietnam, etc. These dolphins start off gray and become pink as they mature. Worldwide, the population of the pink dolphins is small. In order to stop the decline trend in the number of Indo-Pacific humpback dolphins and to reduce the species' extinction risk, great efforts are needed to protect them [9].

Apparently, physical dolphin-assisted therapy is not cheap, besides the involvement of professional therapists and dolphin trainers. The therapy may not be always available. For example, a dolphin may not be ready for therapy sessions depending on its physical conditions [8]. It may also bring risks to children during physical dolphin encounters. With the technology advancement in 3D animations and virtual reality (VR), serious games bring the possibility to model different types of real-world scenarios into virtual learning environments for the purpose of educating or problem solving. The 3D models of dolphins and training pools can be built and projected with high-quality animations and audios to mimic the real dolphins swimming in the pools. Such VR serious game for pink dolphins can be used to replace the physical dolphins in the therapies [3]. It can address the challenges encountered by the physical dolphin-assisted therapy for children with ASD.

This chapter depicts the design process and modeling techniques for virtual pink dolphin serious game. This involves ideations, 3D modeling, texturing, animation, and case analysis.

1.2 Objectives and Scopes

This research work is to explore and investigate the modeling of virtual pink dolphins and relevant virtual game objects. It aims to interact with virtual pink dolphins for serious game design by developing 3D visualization, simulation, and interactive techniques. VR technology equips the virtual pink dolphin serious game with multi-sensorial channels, including visual, audio, and tactile for human–dolphin communication. Research will be directed towards better understanding of the anatomy, biomechanics, and social behaviors of pink dolphins as well as their living environments.

The ultimate goal of this research work is to develop a virtual pink dolphin-assisted therapy (VPDAT) and human-mediated learning for children with ASD. Hopefully, the developed virtual pink dolphin games can be used for the real pink dolphins assisted therapy. The 3D virtual pink dolphins and virtual game objects can be the building blocks to develop an educational program using the latest technologies to raise the public awareness for protecting the vulnerable species.

The virtual pink dolphins work can be also directed towards a long-term development of a unique entertainment program, which can be served as a potential tourism attraction value-added to Singapore tourism industry.

1.3 Organizations of the Chapter

Tools and softwares, which are used in the development of virtual models and objects for this chapter are introduced in Sect. 2. Section 3 describes the design details of 3D modeling of the virtual pink dolphin and a virtual lagoon. The animations and path controls of the virtual pink dolphins and other virtual objects are presented in Sect. 4. Section 5 concludes this chapter and discusses future development.

2 Design Software Tools Used in This Work

2.1 Autodesk 3ds MAX

Autodesk 3ds MAX is a tool for modeling, animation, global illumination, dynamic simulation, and rendering with flexible plugin architecture running on different platforms. 3ds MAX tool has been used in movie making, computer gaming, architectural, and engineering industries. It has the built-in scripting language, MAXScript, which can be used to create plug-in modules [1]. There are different sets of tools available as plug-ins in the 3ds MAX for different applications. Autodesk 3ds MAX has some unique features and primitive shapes to build models and optimize the models for further development. The game design can use polygon modeling to control individual polygons in models. This feature brings convenience of optimizing 3D models.

2.2 Adobe Photoshop

As a popular image editing software tool, Adobe Photoshop is used to draw basis of 2D designs of models of the virtual dolphin and the lagoon, before building their 3D models.

2.3 Zbrush

Zbrush is a digital sculpting and painting software tool, which can help designers to create models and illustrations quickly [14]. Zbrush allows designers to use customizable brushes to shape, texture, and paint virtual clay in a real-time environment with instant feedback provided. High-resolution models with tens of millions of polygons can be easily handled using Zbrush. A quality base mesh leads to a great finished model, which can be generated by Zbrush modules Sculptris Pro, Dynamesh, Zmodeler, etc. The mesh details can be exported as normal mapping or displacement map in 3D computer graphics.

3 Design Details of 3D Modeling

3.1 Design of Virtual Pink Dolphin Models

The mood of designers has a great influence on the draft drawings and the final delivered outputs in the concept design process. Before the start of design, a lot of reference dolphin pictures in real world have been studied. Then the style and size of virtual pink dolphins are determined with the concept dolphins are drawn in 2D images. Figure 1 shows the concept design of virtual dolphins.

High Polygon Model of Dolphins

The 3D models can be created using software tools such as Autodesk Maya, 3ds MAX, Zbrush, or other engraving software. Each tool has its own unique features, advantages, and disadvantages. Models, as vector based in 3ds MAX, are built by points being placed in a 3D space, i.e., X-axis, Y-axis, and Z-axis. The manipulation of the constructed models is done by moving the points in 3D spaces. Zbrush behaves like painting displacement map tool in real time, with a painterly approach to manipulating 3D data. Using 3ds MAX tool needs to pay more attention to the way

Fig. 1 Dolphin concept design

of alignment, while Zbrush tool makes much more surfaces when creating models. Autodesk 3ds MAX is selected to create high polygon models as building blocks. The high polygon model of dolphins created is shown in Fig. 2a.

Intermediate Polygon Model of Dolphins

For the required details of a high-quality dolphin, the high polygon model is made from the intermediate polygon model, to ensure that part of engraving goes simple line and easy to carve. Figure 2b shows the intermediate polygon dolphin model.

Low Polygon Model of Dolphins

Low polygon models can be baked from the high polygon models in 3ds MAX tool. The low polygon models should match the high polygon models to help designers make UVW map. The UVW coordinates are similar to the XYZ coordinates, where the U-, V-, and W-axes of a bitmap correspond to the X-, Y-, and Z-axes, respectively. The W-axis is used for procedural maps. There is a module in 3ds MAX, *UVW Map modifier,* provides mapped and procedural materials appearing on an object's surface. It offers a convenient and quick way to control mapping coordinates of a model object.

Low polygon models can be also modified from the intermediate polygon models. In design process, attentions should be paid to the overall outline, which is the most important part in low polygon model. The low polygon model should not be fragmented, in order to make it easy to keep modeling. In this design of virtual pink dolphins, the flippers, tail, and head with body are combined into a single part of the model. The low polygon dolphin model is shown in Fig. 2c.

UVW Map of the Dolphin Models

In the game design, UVW map plays an important role in modeling. Only when the UVW map is guaranteed, the texture of the models can be made well accordingly.

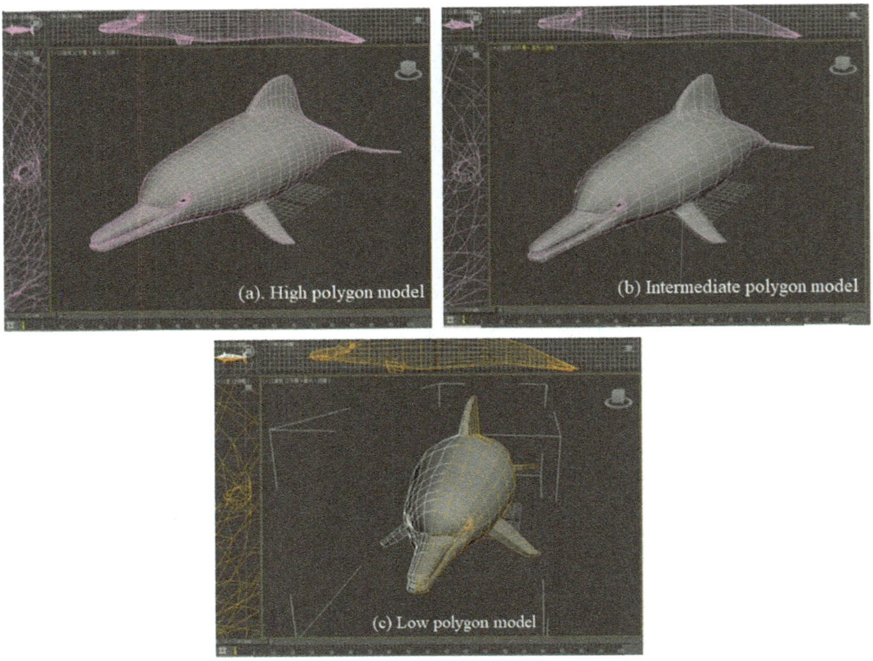

Fig. 2 Created dolphin models, **a** high polygon, **b** intermediate polygon, **c** low polygon

The UVW map can be designed by various software tools depending on the design facilities. The UVW map of the designed dolphin models for this work is shown in Fig. 3a.

Texturing of the Dolphin Models

In the next design step, skins of the dolphin models are drawn using Acrobat Photoshop on the UVW map of dolphin models, shown in Fig. 3b.

After that, the dolphin models are loaded into 3ds MAX with the final effect shown Fig. 3c.

Modifications and Fine Tuning

With the prototype of virtual dolphin models drawn and viewed in above-mentioned design steps, designers can then check if the built model is suitable to the serious game; if the number of surfaces match the design requirements; if polygon is in legality; if the texture is twisted by triangular poly; and so on. This step is also important in the game modeling. If necessary, designers can perform some minor modifications and fine-tuning on the 3D models at this step. Two examples of minor modifications on the dolphin models are shown in Fig. 4.

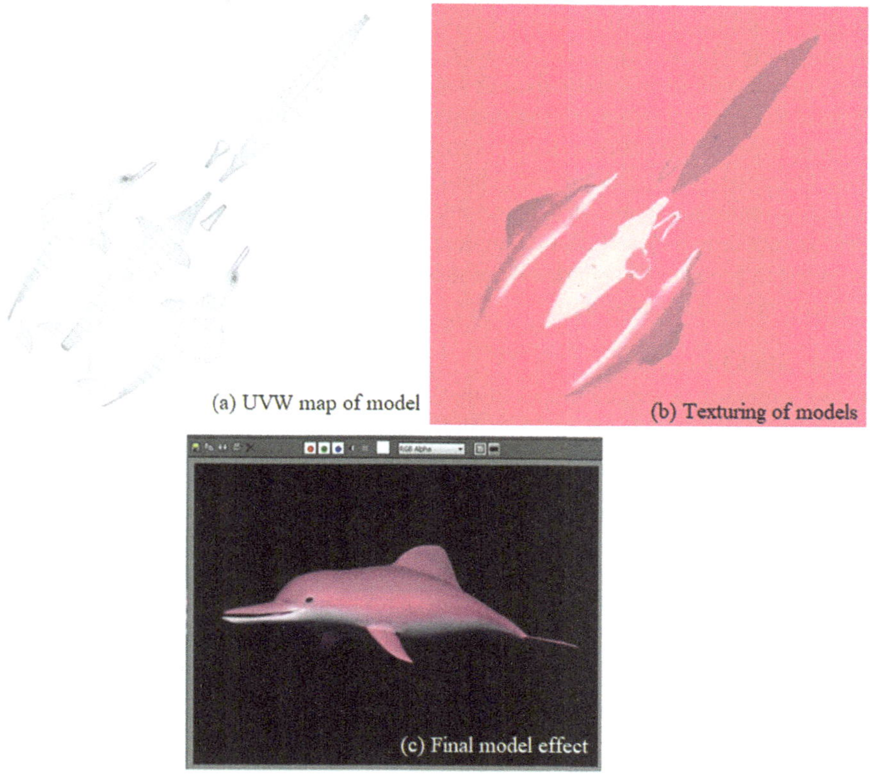

Fig. 3 Dolphin models design, **a** UVW map, **b** texturing, **c** final effect

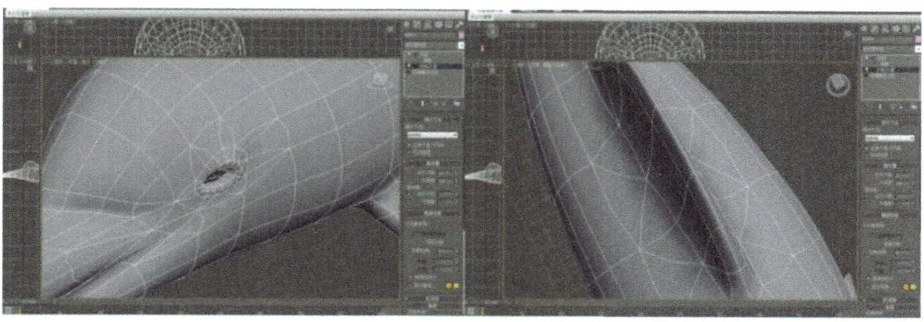

Fig. 4 Two examples of minor modifications on the models

3.2 Design of Virtual Lagoon Models

In order to create a virtual lagoon where virtual pink dolphins can swim and interact with children, it is expected to have a cute design to attract the attentions of children.

We have tried out many different concepts in the work. Finally, this concept design shown in Fig. 5 is selected. This concept has a cute design with a turtle carrying a huge lagoon with some classical castles at the side of the lagoon. Pink dolphins can swim in the virtual lagoon in a cartoon style, which is loved by children.

To better craft the concept design, some pictures of real turtles are used for reference. Once the overall design and general sense have been fixed, the design details and some important items need be considered. In the concept design in Fig. 5, some parts are not drawn clearly. For example, the details of the houses on the lagoon are too small to be seen. Design details of the important items need to be implemented carefully in the designing steps.

Simple Model Creation
At the beginning, a simple model of the lagoon design is built as shown in Fig. 6, which can represent the proportion and structure of the model according to the concept design. This is a common method used in the game development process.

Building of High Polygon Model
In 3D modeling, normal map is very crucial through which details of the body on the low polygon model can be shown. The structures in this map are from the high polygon model being built, shown in Fig. 7. It is required for the high polygon model to show an overall structure and reflect details of the model like the thickness, depth, texture, and the feeling. The process file needs to be saved during the processing of making high polygon model. Each component needs to be stored separately as a simple model, detail model and final model.

Detail models are the ones well organized without problem in scales. This type of detail models can easily make changes if the models have some errors, while final models can directly show the effect of high polygon models.

Fig. 5 Concept design of a lagoon carried by a turtle

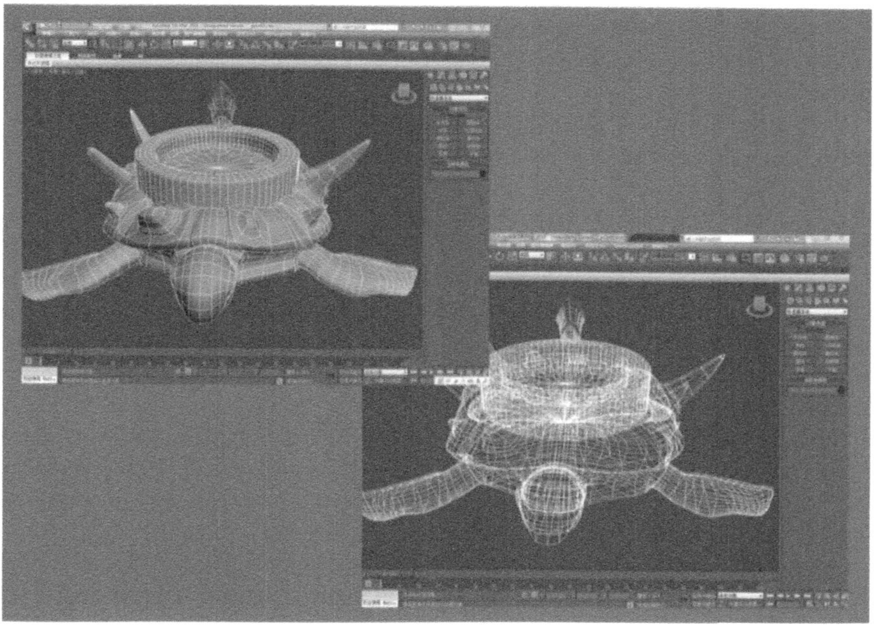

Fig. 6 Simple model of the lagoon design

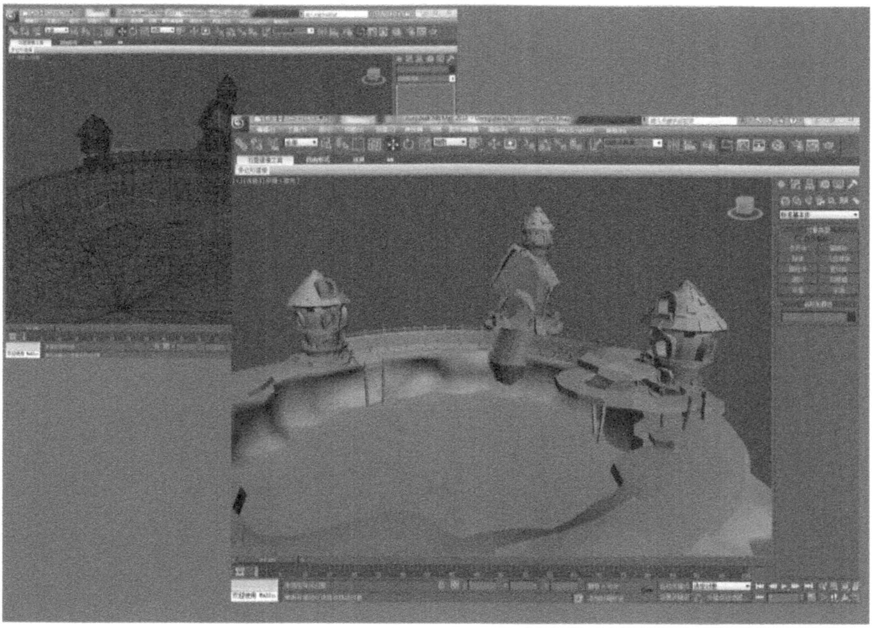

Fig. 7 High polygon model of the lagoon design

The simple polygon models are useful to make the low polygon model. If simple polygon models are well structured, they can be directly modified to low polygon models.

During the processing of high polygon models, the specifications and requirements of low polygon models need to be considered.

Creating Low Polygon Model and UV Map

With the saved files after the process of making high polygon models, it is very convenient to build the low polygon models. If the wiring is generally smooth and regular without number of wasted surfaces, the effective approach is to build the models first with all the surfaces being useful. After completion of the models' building, the number of surfaces can be decreased or increased according to the design requirements.

The low polygon lagoon model is shown in Fig. 8. In the designing of low polygon models, the characteristics of normal map need to be understood. It is essential to know what kinds of details need to be presented in the low polygon models and what kinds of details can be expressed well. Normally, the structures, which can be seen from different perspective views should be made in low polygon models. Some examples of structural details that presented in the low polygon lagoon model are shown in Fig. 9a, b.

The cracks in the wall, brick structures, and the small dirt structures can be expressed by baking normal map in Autodesk 3ds MAX or other normal mapping

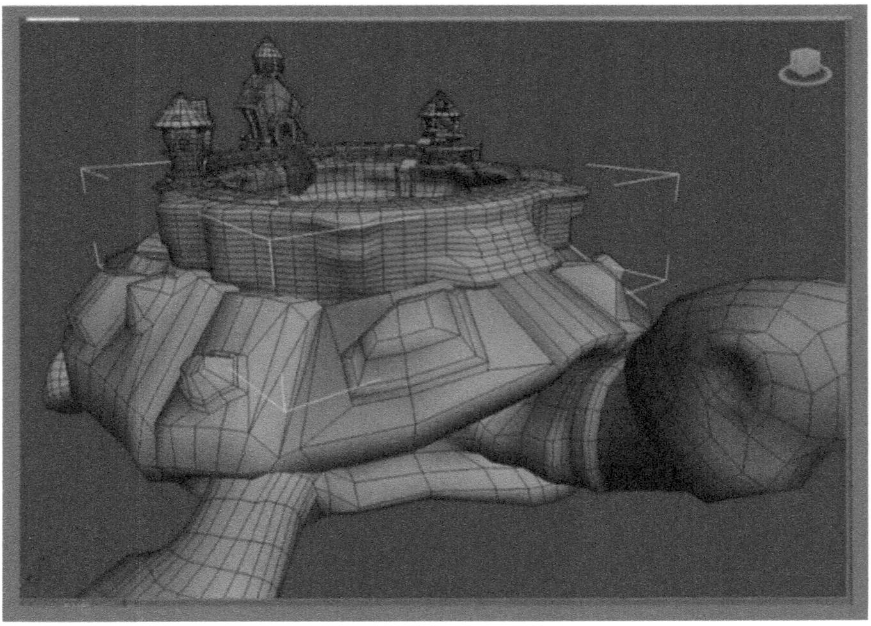

Fig. 8 Low polygon lagoon model

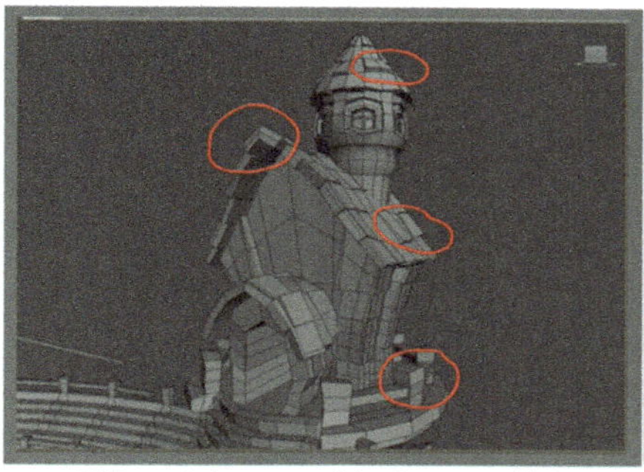

(a) Example 1 - structural details presented in low polygon lagoon model

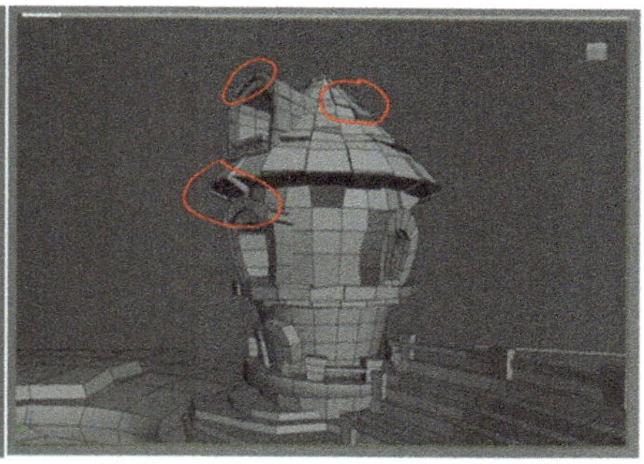

(b) Example 2 - structural details presented in low polygon lagoon model

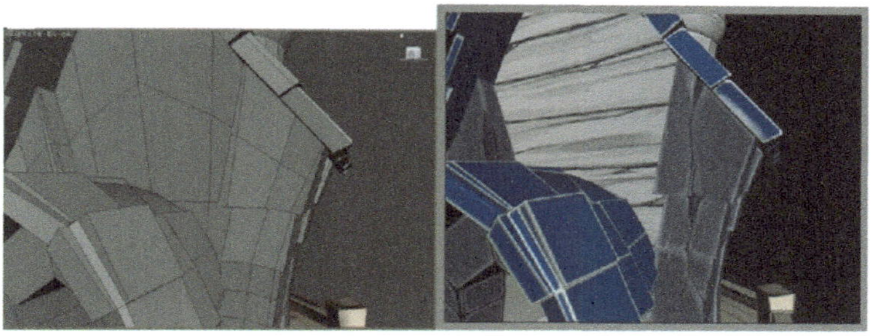

(c) Example – final details presented in low polygon lagoon model

Fig. 9 Examples of structural details presented in low polygon lagoon model

software tools. In this work, Fig. 9c shows an example of the final normal map of low polygon modeling of the lagoon model.

UV Map Layout of Lagoon

UV mapping is a 3D modeling process of creating explicit UVs for a surface mesh, where a 2D image is projected onto a 3D model surface with texture mapping. When exhibiting UV map, it is good to utilize the space well in UV edit box. The step of UV mapping is a critical skill to ensure the design accuracy and realistic textures on polygonal surfaces. It is important for the game development. The overview of the UV mapping of the lagoon model is shown in Fig. 10.

Diffuse Map of Lagoon

This process requires game designers to have a good understanding of sketch and good sense of color, as many highlights and shadows are drawn stroke by stroke manually. It usually takes a lot time in this process to draw for high-resolution pictures with 2048×2048 pixels. The suitable materials and damage effects need to be drawn on every diffuse map according to different objects and different materials. The texturing of diffuse map is shown in Fig. 11.

At the end of the design process, the lagoon models will be loaded into Autodesk 3ds MAX. The final effects of the designed lagoon model are shown in Fig. 12.

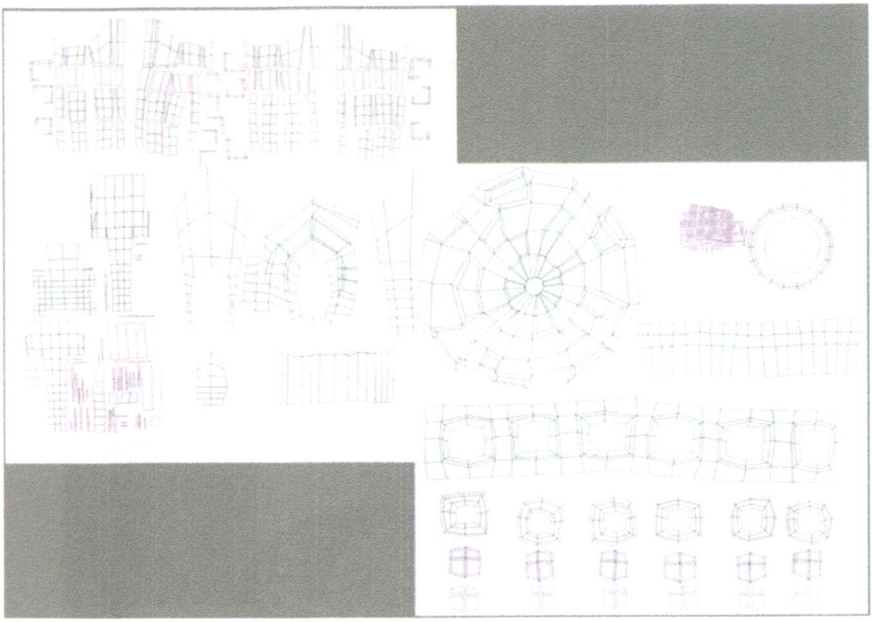

Fig. 10 UV map layout overview

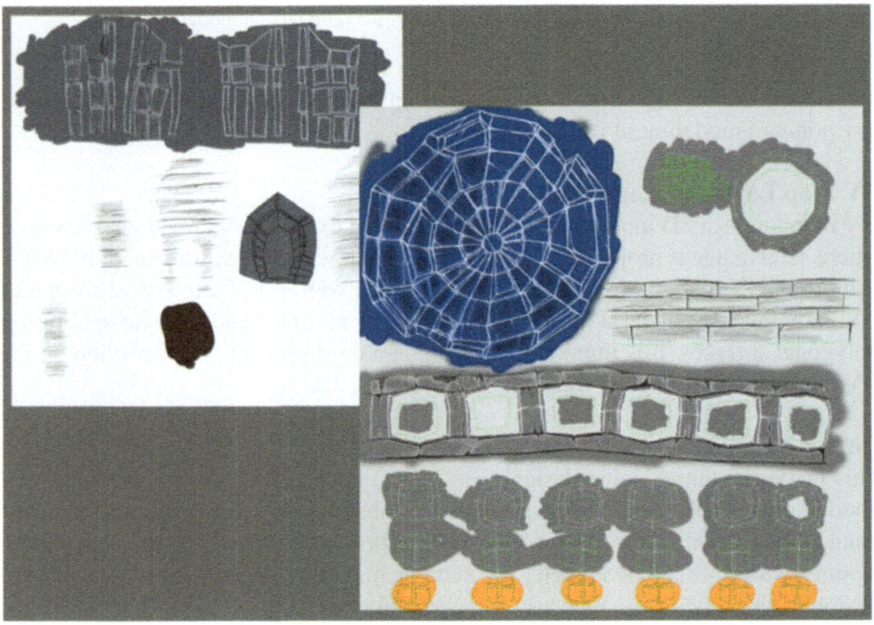

Fig. 11 Texturing of the lagoon model

Fig. 12 Final effects of designed lagoon model

The developing process of lagoon models is relatively long in this work. It requires the designers to equip various knowledge throughout drawing, modeling, and creativeness. The produced lagoon model fulfills the design requirements of the targeted serious game. But, there are still a few flaws in the models, which should be fixed. For example, the high polygon model of the lagoon can be more detailed, and the UV map can be more reasonably distributed.

3.3 *Other Virtual Objects in the Game*

In order to make the virtual pink dolphin serious game more interesting, other virtual objects or elements can be created in the game, using the same design process. In this research work, an octopus emcee is introduced in the virtual stage of the serious game. The virtual octopus can play and interact with the virtual dolphin using its paws. Figure 13 shows the virtual octopus emcee holding a hula hoop. The reason why octopus is chosen instead of other animals is that an octopus has eight paws to hold more virtual objects than other sea animals. Furthermore, the appearance of octopuses is cute and loved by children.

Fig. 13 Virtual octopus model in the Game

4 Animation of Dolphins

In the targeted VR serious game, dolphins will swim inside the lagoon and interact with game players through gesture control. Such activities are implemented with animations and path control for each modeled 3D object.

4.1 Dolphin Skeleton Creation

Before creating the skeleton, the overall bone structure of dolphins has been studied with images found in literature. In this game development, the skeleton of dolphin models has been simplified, in order to reduce the design complexity. The skeleton of the dolphin models is made up of head, flippers, and body bones. All the three parts of the bones are jointed together with link function, shown in Fig. 14.

4.2 Path Control

The path of animation is to control objects moving through the trajectory in virtual spaces. It is very useful animation control function, besides the function of control key frame. Using the path control can make the dolphin models doing more complex displacement. The path control approach is much easier than that of using key frames.

The path control for dolphin models' animation is shown in Fig. 15. A new path of animation can be added by clicking the button of Add Path. It lets the dolphin model attach to the created circle and perform the movement along this path.

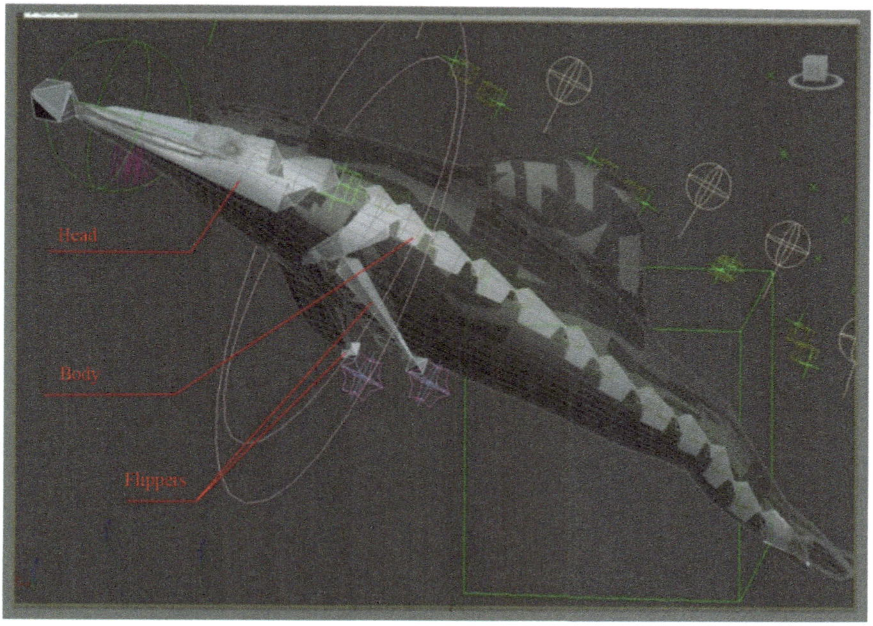

Fig. 14 Skeleton of virtual dolphin models

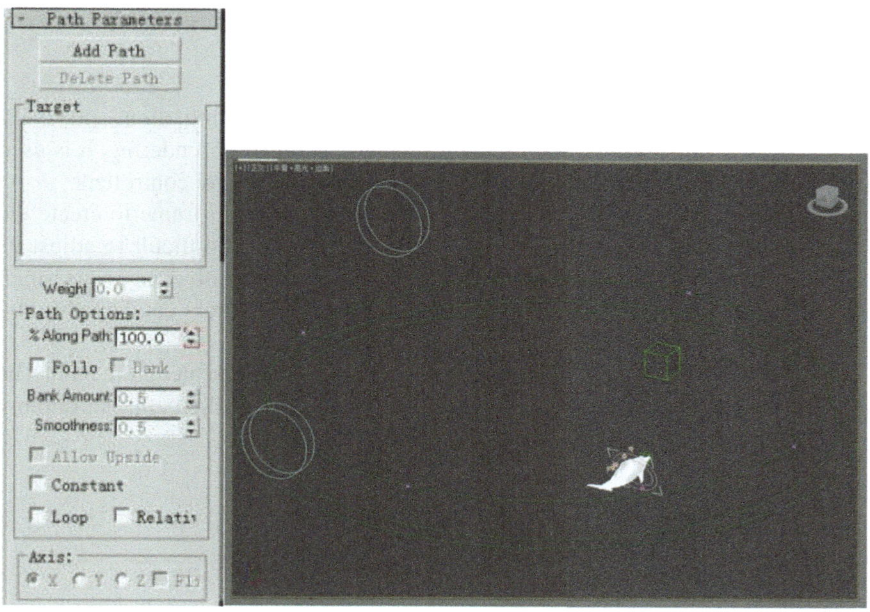

Fig. 15 Path of animation for dolphin models

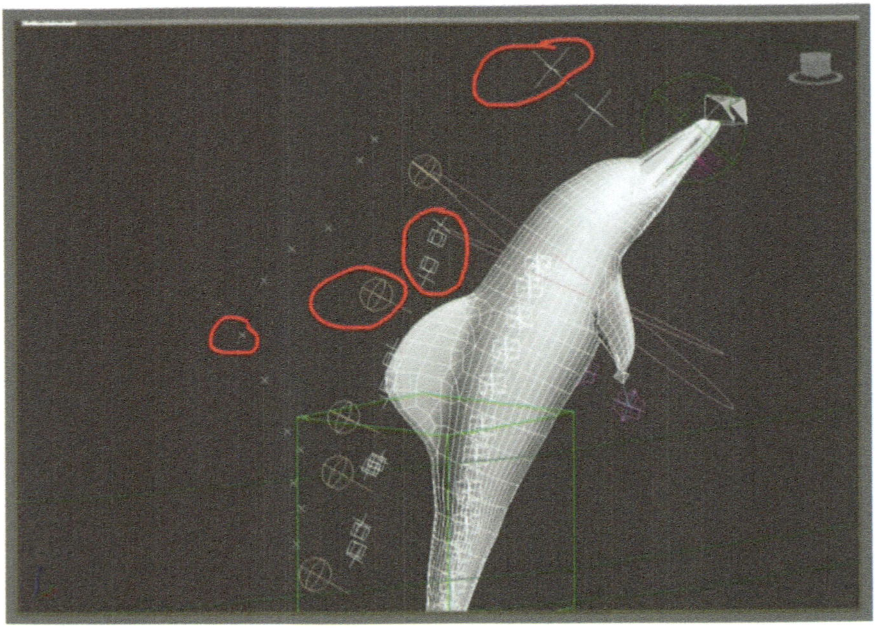

Fig. 16 Dummy of dolphin models

4.3 Making Dummy of Dolphins

The main role of the dummy function is to create complicated animations and levels. As the dummy function in the game design is not visible after rendering, it is used to make offset joints between the object connectors and the level controllers.

Key frames can be adjusted by setting the location of each frame to create the animation of dolphin moving along the paths. But it is very difficult to adjust its height or movement later. Many frames need to be adjusted with even a very small movement. With the usages of the dummy function, this problem can be solved by decomposing the objects movement.

Shown in Fig. 16, the dolphin model can move up and down along the path. The height of dolphin jump can simply be adjusted by changing the path that dummy created.

4.4 Curve Editor

The curve editor can be used to edit the movements of dolphin model animations. The editing is performed to change the properties of a curve. Designers can select the key points on the curve and then right click the mouse button to show the properties functions. A sample curve editor of the animation as shown in Fig. 17.

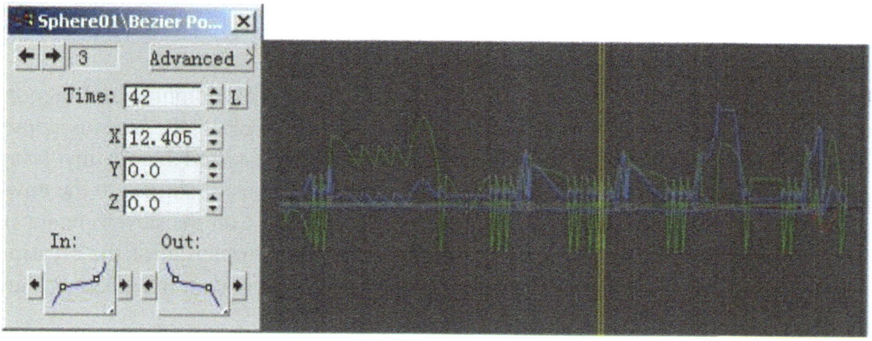

Fig. 17 Sample curve editor of animation

Using the curve editor, the moving speed of the animation can be also edited. For example, the acceleration of gravity must be considered when the virtual dolphin did a jump movement. The properties of the acceleration cure can be modify in the curve editor, shown in Fig. 18.

Using the above-mentioned design processes for the animations in this section, each virtual object including pink dolphins, octopus, and lagoon on the back of the turtle can move around individually in the serious game. It makes the game realistic and attrative to children.

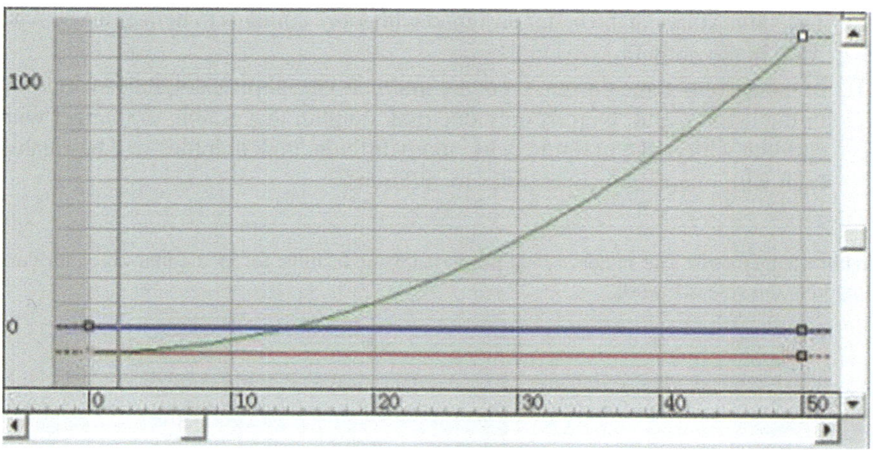

Fig. 18 Curve editor to modify acceleration of movements

5 Conclusions

This chapter details the design processes of the virtual pink dolphins and lagoon. Various 3D virtual models are created as the building block of the serious game. The aim of this VR serious game is to simulate the dolphin-assisted therapy involving virtual pink dolphins, octopus emcee, lagoon, and some houses, similar to the environment of underwater parks. This chapter mainly focuses on the development of the 3D virtual models of these game objects. After the completion of the creation of these virtual models and their attached animation properties, in this chapter, they will be used in the serious game.

To develop 3D games, designers need to have good understanding and good skills of 3D modeling, graphical design, game engines, features of various design tools. Certain design experiences are required to develop these virtual models, which match design requirements of the serious game in the next phase.

In future development of the work, several further enhancements are expected.

(1) The current lagoon model and design are suitable to the dimension of a single display screen. It is suitable to the serious game running on the computer with a single player. If the serious game is running in a 3D CAVE with surrounding screens, for example, a 3D Immersive Room at Institute for Media Innovation with five display panel span in 320-degree, the built lagoon model is too small to fit in. In this case, the 3D lagoon modeling need to be re-designed in order to fit in the multiple surrounding screens.
(2) The game story can be better designed with more contents to help children with ASD in the dolphin-assisted therapy.
(3) In the current game design, there are multiple virtual pink dolphins swimming in the lagoon. But, there is only one pink dolphin that is able to interact with children with ASD. In future work, more different pink dolphins will be created each with interaction capabilities to children.

Acknowledgements The authors would like to thank the Institute for Media Innovation for their support to this research work.

References

1. Autodesk Inc.: Getting started in 3ds MAX (2018). https://area.autodesk.com/tutorials/series/getting-started-in-3ds-max-2018/. Accessed 30 July 2020
2. Brensing, K., Linke, K.: Behavior of dolphins towards adults and children during swim-with-dolphin programs and towards children with disabilities during therapy sessions. Anthrozoos Multidiscip. J. Interact. People Anim. **16**(4), 315–331 (2003). https://doi.org/10.2752/089279303786992035
3. Cai, Y., Chia, K., Thalmann, D., Kee, K., Zheng, J., Thalmann, N.: Design and development of a virtual dolphinarium for children with autism. IEEE Trans. Neural Syst. Rehabil. Eng. **21**(2), 208–217 (2013). https://doi.org/10.1109/TNSRE.2013.2240700
4. Cai, Y., Chiew, R., Nay, Z.T., Indhumathi, C., Huang, L.: Design and development of VR learning environments for children with ASD. Interact. Learn. Environ. **25**(8), 1098–1109 (2017). https://doi.org/10.1080/10494820.2017.1282877
5. Cai, Y., Goei, S.L., Trooster, W. (eds.): Simulation and Serious Games for Education. Springer, Singapore (2016)
6. Candelieri, I.: Healing and caring in dolphin-assisted therapy: criticisms of effectiveness and ethical issues. Gestalt. Theor. **40**(3), 323–335 (2018). https://doi.org/10.2478/gth-2018-0024
7. Fiksdal, B., Houlihan, D., Barnes, A.: Dolphin-assisted therapy: claims versus evidence. Autism. Res. Treat. (2012). https://doi.org/10.1155/2012/839792
8. Griffioen, R., van der Steen, S., Cox, R.F.A., Verheggen, T., Enders-Slegers, M.J.: Verbal interactional synchronization between therapist and children with autism spectrum disorder during dolphin assisted therapy: five case studies. Animals (Basel) **9**(10), 716 (2019). https://doi.org/10.3390/ani9100716
9. Jefferson, T., Smith, B.: Re-assessment of the conservation status of the indo-pacific humpback dolphin (Sousa chinensis) using the IUCN red list criteria. Adv. Mar. Biol. **73**, 1–26 (2015). https://doi.org/10.1016/bs.amb.2015.04.002
10. Kreivinienė, B., Kleiva, Ž.: Subjective approach towards the welfare understanding in the dolphin assisted therapy: experiences of families in pilot research. Soc. Welfare Interdiscipl. Approach **7**(1), 142–157 (2017). https://doi.org/10.21277/sw.v1i7.291
11. Li, S.: Humpback dolphins at risk of extinction. Science **367**(6484), 1313–1314 (2020). https://doi.org/10.1126/science.abb5744
12. Lu, A., Chan, S., Cai, Y., Huang, L., Nay, Z.T., Goei, S.L.: Learning through VR gaming with virtual pink dolphins for children with ASD. Interact. Learn. Environ. **26**(6), 718–729 (2017). https://doi.org/10.1080/10494820.2017.1399149
13. O'Haire, M.: Research on animal-assisted intervention and autism spectrum disorder, 2012–2015. Appl. Dev. Sci. **21**(3), 200–216 (2017). https://doi.org/10.1080/10888691.2016.1243988
14. Pixologic Inc.: ZBrush at a glance (2020). http://pixologic.com/features/about-zbrush.php. Accessed 30 July 2020
15. Salgueiro, E., Nunes, L., Barros, A., Maroco, J., Salgueiro, A., dos Santos, M.: Effects of a dolphin interaction program on children with autism spectrum disorders: an exploratory research. BMC Res. Notes **5**(1), 199 (2012). https://doi.org/10.1186/1756-0500-5-199

Serious Game Design for Virtual Dolphin-Assisted Learning

Weiliang Ryan Liu, Qi Cao, and Yiyu Cai

Abstract Virtual dolphin-assisted therapy (DAT) can be used to treat children with autism spectrum disorder (ASD), where interactions with virtual dolphins can be enabled by Virtual Reality (VR) technology. With the aid of VR, it is possible to reproduce an immersive environment with virtual objects to replicate the real-world DAT interactions. Using VR to create human–virtual dolphin interactions in an immersive setting allows DAT to be carried out virtually, protecting both participants, children and dolphins, from any harm or infections. The aim of recreating a DAT VR environment is also to make the treatment cheaper and more readily accessible to the masses of children with ASD. It can control the treatment environment and ensure the safety of the participants and those parties involved. A VR serious game, 3D Virtual Pink Dolphin is developed in Nanyang Technological University, which is running on both PC with single display screen and a 3D immersive room with 320-degree curved screens. The experiment has been conducted with children from a special needs school in Singapore.

Keywords VR serious game · Dolphin-assisted therapy · 3D display · Special needs education

W. R. Liu · Y. Cai (✉)
School of Mechanical and Aerospace Engineering, Nanyang Technological University, Singapore 639798, Singapore
e-mail: myycai@ntu.edu.sg

Q. Cao
School of Computing Science, University of Glasgow, Singapore Campus, Singapore 567739, Singapore
e-mail: qi.cao@glasgow.ac.uk

Y. Cai and Q. Cao (eds.), *When VR Serious Games Meet Special Needs Education*, Gaming Media and Social Effects, https://doi.org/10.1007/978-981-33-6942-9_6

1 Introduction

1.1 Background

There are many therapeutic methods in the literature for improving outcome or facilitating recovery of autism spectrum disorder (ASD) [9]. Virtual social skills interventions have been received attentions attributing success to interactions of avatars controlled by therapists and patients in a virtual environment [10]. There are different types of virtual avatars for therapies and treatments, such as animated avatar with three-dimensional (3D) appearance, cartoon avatar with two-dimensional cartoon character, human-like avatar, and animal avatar [7]. Virtual reality (VR) serious games mimicing the dolphin-assisted therapy (DAT) is one of such methods with animal avatar utilizing sensor devices, 3D modeling and graphical rendering to create immersive virtual environment [1]. Children with ASD can learn social skills and safety skills through interacting with virtual dolphins using hand gestures. Such immersive VR enables the possibilities of manipulating visual, audial, and sensory signals to bring a treatment for participants in a controlled environment [6]. It can improve communication and learning skills of children with ASD [5]. Virtual 3D environment can encourage interactivity and motivation of players as well as improve their knowledge and enjoyment in the learning process [11]. Often such research is of multidisciplinary nature involving special needs educators, researchers, engineer, and developers. An international research collaboration among Singapore, the Netherlands, and China on VR technology to improve children with ASD in their learning has been reported by Cai et al. [2]. Under this cross-continents research program, researchers, principals, teachers, and students from multiple countries work jointly to develop special needs learning technology and contents. An initial exploring and design work of a virtual dolphinarium for children with ASD is presented by Cai et al. [1]. Lu et al. [8] describe their works with preliminary evaluations for children with ASD interacting with virtual pink dolphins, to learn direction following, psychomotor skills and hand–eye coordination.

The VR serious games can be running on computers with graphics processing unit (GPU) or mobile devices. The contents of the serious games could be projected and displayed on 3D curved surrounding screen, TV screens, head-mounted display (HMD), computer monitors, or screens of mobile devices. Different types of screens provide different user experiences in 3D with the VR serious games. The HMD and 3D curved screen provide surrounding VR effects [4]. The HMD is considered as high-end immersive approach [12]. The 3D curved screen projects graphic contents and virtual environment on seamless curved wall screens that surrounds players as immersion effects. Through 3D glasses, players can see real objects and their own actions in 3D environment, which enables them to get real-time visual feedback in the learning process of serious gaming [3].

1.2 Objectives of This Chapter

This chapter introduces the implementation details of the virtual dolphin serious game and the experimental study with a special needs school in Singapore. The design of objectives in the dolphin-assisted learning serious game is to provide a platform and a tool for training children with ASD. The virtual pink dolphin serious game encourages children with ASD to interact with virtual objects in surrounding environment. It can also encourage learning of basic daily skills, which are easy to normal children but challenging for children with ASD.

1.3 Scopes

The scopes of this research work are included as follows:

(1) Get children with ASD to play the VR pink dolphin serious game.
(2) Collect feedbacks from parents and teachers after the plays of (the VR serious game by children with ASD).
(3) Collect information on how special needs school uses interactive VR technology to help children with ASD.
(4) Use collected feedback to further improve the pink dolphin serious game, which can be used as a tool for rehabilitation of children with ASD.

1.4 Organizations of the Chapter

In this chapter, Sect. 2 depicts information about the VR pink dolphin serious game, such as the equipment used, the game play, and the game development. The experiment of virtual pink dolphin serious game has been conducted with children from Asian Women's Welfare Association (AWWA) school, a special needs school in Singapore. The experiment setup, procedures, and feedback are presented in Sect. 3. The observations and feedbacks of the experiment are discussed in Sect. 4. Section 5 concludes this chapter.

2 Operations of Virtual Pink Dolphin Serious Game

2.1 Equipment in 3D Immersive Room

The developed virtual pink dolphin serious game can be operated on both single display screen and 3D curved surrounding screen. In this work, the experiment of the virtual pink dolphin serious game will be conducted in the 3D curved screen, i.e., an immersive room located at the Institute for Media Innovation (IMI) in Nanyang Technological University (NTU), Singapore. The immersive room has a dome shape wall with 320-degree screen. The graphic contents are projected by five 3D projectors located at the top of the room.

Players need to wear active 3D glasses in order to see the 3D stereographic contents displayed on the curved screen. There are multiple motion detection devices surrounding the 3D immersive room to capture gestures of players, which is used for players to interact with virtual pink dolphin in the VR serious game.

Other equipments in the 3D immersive room include the speakers to play game audios and a life jacket, which is fitted with sensors and water spray. The audios played through speakers act as audio instructions or feedback in the VR serious game. Players wear the life jacket while playing the pink dolphin serious game. The water sprays and sensors in the life jacket act as a fourth element over the 3D serious game, by spraying water on players when the virtual pink dolphin does a trick or splash water on players.

2.2 Two Modes in Virtual Pink Dolphin Serious Game

There are two game modes in the virtual pink dolphin serious game developed, which players can choose from the main game menu. The first game mode is the free interaction. The second one is the goal orientated game. For both game modes, virtual pink dolphins can react to hand gestures given by players. The hand gestures will be captured by the motion detection devices surrounding the 3D immersive room, in such way the interactions of players and the virtual pink dolphins are achieved.

Free Interaction Game Mode

In the free interaction game mode, players can freely interact with the virtual pink dolphin without any preset goals or objectives, as shown in Fig. 1. The game mode allows the players to act as a dolphin trainer with command gestures, like dolphin trainers at Sentosa's Underwater World, Singapore. Players can use controlled command gestures shown in Fig. 2, to interact with virtual pink dolphins in the game.

The command gestures given by players are similar to those hand gestures used by real dolphin trainers in Sentosa's Underwater World. By having similar hand gestures, it adds a factor of realism to the game play, which can be helpful when the game is to be used as a virtual DAT tool for children with ASD.

Players first do the game calibrations with motion detection devices by waving both hands shown in Fig. 2a. The game players can interact with the left dolphin in a group of dolphins using the command gesture shown in Fig. 2b. The dolphins on the right side and in the middle can be interacted by giving the command gestures shown in Figs. 2c, d, respectively. The command gesture in Fig. 2e instructs the virtual dolphins to spray water to the game players through the life jacket.

Fig. 1 Interacting in the free interaction game mode

Fig. 2 Command gestures in the free interaction game mode

By having this free interaction game mode with the virtual pink dolphin, players are able to learn how to control virtual pink dolphins. This game mode provides players the chance to experience the feeling of being real-life dolphin trainers virtually.

Goal Orientated Game Mode

In the goal orientated game mode, there is a human avatar shown on the screen, which requires players to perform responding actions according to the instructions given by the human avatar. The human avatar gives both text instructions and corresponding hand gestures to players. Players are required to follow the instructions and imitate these hand command gestures. These gestures in the game mode are shown in Fig. 3. By correctly imitating each hand gesture, the virtual pink dolphin would collect a green gemstone on the screen. The objective of the goal orientated game mode is for players to complete the serious game by collecting all the gemstones step by step. In the orientated game mode, players must complete all instructions in order to move on to the next stage. This game mode teaches children with ASD to follow instructions with the rewards of seeing the virtual dolphin and reacting to the command gestures.

The hand gestures given by the human avatar are similar to some of the hand gestures used by the real dolphin trainers at Sentosa's Underwater World, Singapore, as shown in Fig. 3.

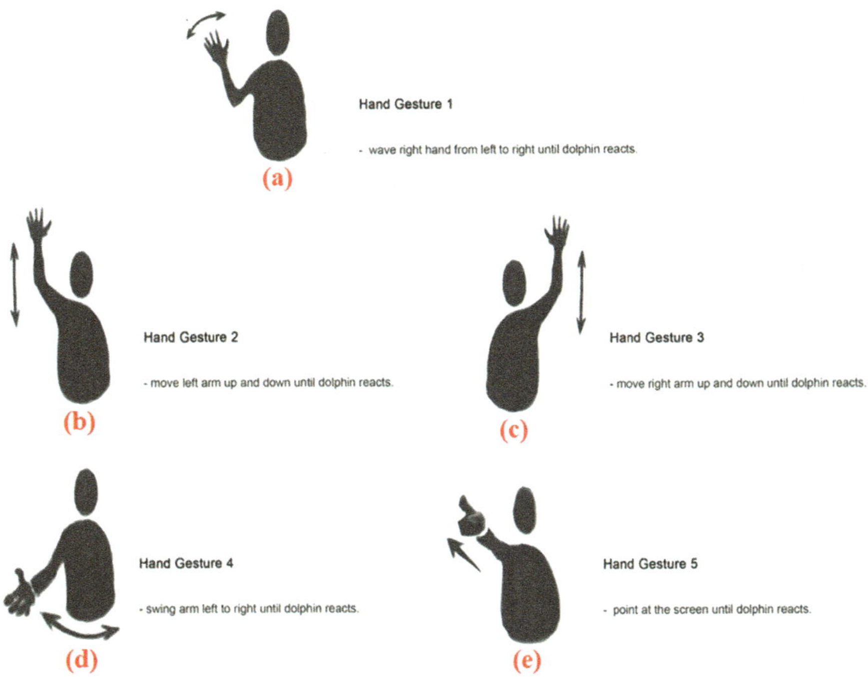

Fig. 3 Command gestures in the goal orientated game mode

3 Experiment Setup

3.1 Experiment Purpose

The purpose of conducting the experiment is to evaluate the effectiveness of the pink dolphin serious game as a treatment tool for children with ASD. It is important to carry out actual test with children with ASD to get their reaction, behavior, and feedback during and after the game play. Their reaction and behavior could be highly unpredictable when being exposed to the game, which are crucial to improve the design of the serious game.

The experiment is conducted in parallel sessions with children from AWWA School in Singapore. In order to better teach children with ASD, AWWA School is actively looking for new pedagogy and teaching technology. Currently, AWWA School has two teaching methods conducted in the class. One of them uses conventional classroom teaching, where a teacher with the aid of a teaching assistant teach a class of 5–6 students. The other method is to use supportive tools, where teachers use alternative approaches to encourage learning. Alternative approaches include using a multi-sensory room to relax and calm the children. The multi-sensory room is equipped with soft objects such as cushion, mattress and form structures, light-emitting devices and also a disco ball. When a session is being conducted in the room, there would usually be soothing music playing in the background. The multi-sensory room acts like a reward for the children. Only those children with good behavior are given the privilege of use the room. Hence, this can be used as a method to teach children with ASD to control their behaviors.

Other forms of alternative approaches used are iPad-based, PCs-based, and Xbox Kinect-based games. By using these approaches, teachers find that children with ASD are keener to sit still and learn. Children are much easier to control in such learning approaches.

Interactive VR serious games are one of the new approaches, which AWWA School is actively exploring. In this research work, selected children of AWWA School have participated in the experiment of the virtual pink dolphin serious game.

3.2 Experiment Procedure

There are 12 children with ASD from AWWA School participated in the experiment. These children with ASD are mixed with different autistic conditions: high functioning level, intermediate level, and severe level ASD.

In the parallel experiment sessions, the experimental venue is divided into five different areas, as shown in Fig. 4. Each experimental area allows children with ASD to work their own way to play the pink dolphin serious game. Each child goes through the activities from the experiment area #1 to experiment area #5 in sequential order. The reason for such arrangement is to gradually let children get used to the idea

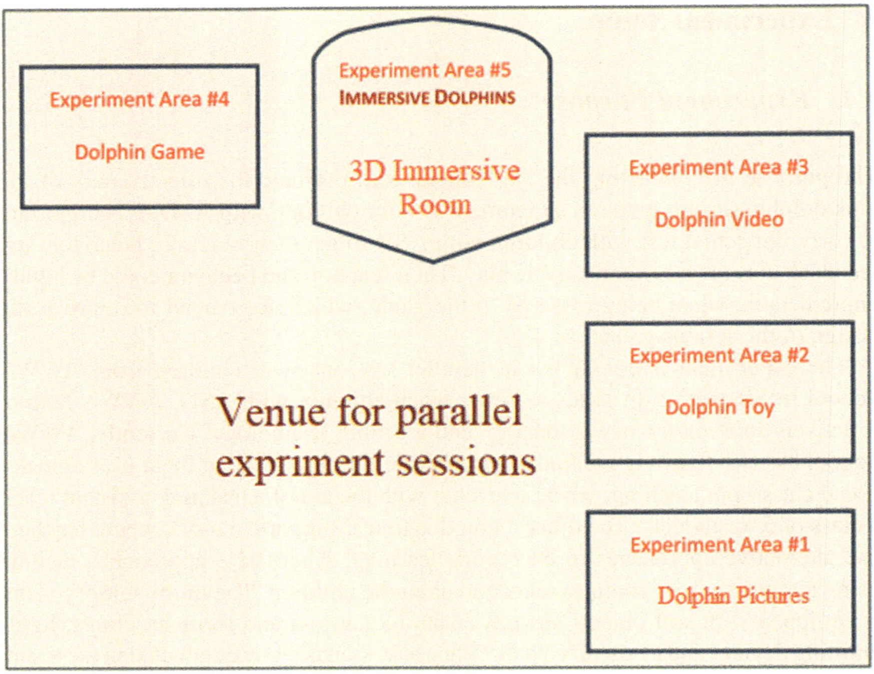

Fig. 4 Layout of experiment venues in parallel sessions

of a virtual pink dolphin interaction, while not exposing the children immediately to the virtual pink dolphins in the immersive room. Without this arrangement, it might be too big of a shock for them to comprehend, and this might contribute to a negative effect to the experiment, as they might not dare to enter the immersive room to interact with the virtual pink dolphins. Major emphasis is placed in the 3D immersive room as one of the major concerns for the children with ASD is that they might feel overwhelmed by the sheer size of the 3D immersive room. For each experiment area, a therapist or special education specialist takes care of the children with ASD, assisted by parents or another teacher.

Experiment Area #1
In the experiment area #1, a therapist or a special education professor works with children by showing the pictures of dolphins and asking the children dolphin related questions, shown in Fig. 5. These questions are to get them interested in dolphins and to prepare the children for meeting the virtual pink dolphins in the 3D immersive room.

Experiment Area #2
For the experiment area #2, the setup is similar to that of the experiment area #1, except for a toy dolphin is used. The children can draw figures of dolphin by themselves, referencing to the dolphin toy, as shown in Fig. 6.

Fig. 5 Activities at the experiment area #1

Fig. 6 Activities at the experiment area #2

Experiment Area #3

In the experiment area #3, a pre-recorded video of a real dolphin trainer from Sentosa's Underwater World is shown to the children. In the video, it shows how the trainer interacts with the real dolphins in the dolphin lagoon, using hand gestures as one of the methods to communicate with the real dolphins, shown in Fig. 7. Some of the hand gestures used in the virtual pink dolphin serious game are similar to those used by the trainer in the video. The activity at the experiment area #3 is to allow the children to link the hand gestures used in the serious game to those of the real dolphin trainer.

The sequential activities get children ready and work them up to the interaction with the virtual pink dolphins. They are asked whether they want to be a dolphin trainer like the person in the video. They are also informed that they will have a chance to interact with virtual pink dolphins later in the experiment area #5.

Experiment Area #4

In the experiment area #4, the children are asked to play the goal orientated game mode of the virtual pink dolphin serious game, shown in Fig. 8. This is to allow them to get used to playing the serious game in a 2D environment. The virtual pink

Fig. 7 Activities at the experiment area #3

dolphin serious game is also in 2D, which is much easier on the children' senses and easier to learn how to interact with the virtual pink dolphins.

Experiment Area #5

The last and final experiment area #5 is in the 3D immersive room. The participants are allowed to freely interact with the virtual pink dolphins shown in Fig. 9. In this experiment area, the children are required to wear 3D glasses to feel the full effect of immersion into the virtual world of the dolphin lagoon.

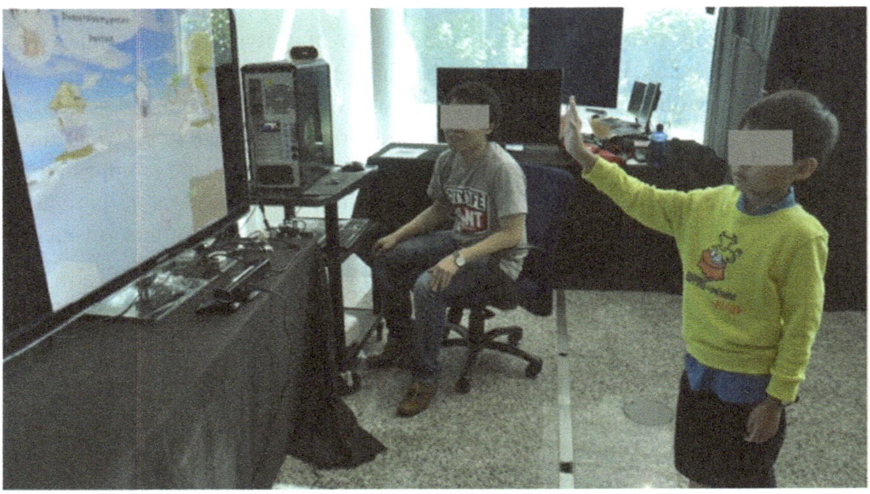

Fig. 8 Activities at the experiment area #4

Fig. 9 Activities at the experiment area #5, in the 3D immersive room

4 Experiment Outcomes and Discussions

The outcomes of the experiment are of mixed results, as the group of children with ASD are with different autistic levels. Some children are of high functioning ASD, while there are a few children with intermediate and severe ASD level. Hence, those children with high-level ASD are able to complete the experiment without any issue.

For those children with severe level ASD, they are either unable to comprehend what is happening in the experiment, or they are even unable to sit through anyone of the experiment area. These children rather choose to run away, or to do their own things. Even with the patient guidance from the therapist, these children are unable to complete any of the experiment area successfully.

The experiment shows that children with high functioning level ASD can benefit from the virtual pink dolphin serious game. It is observed that these children are relatively calmer while interacting with the virtual pink dolphins. They are also more willingly to follow instructions given by the therapist. They are able to express their joy and excitement in the game play.

For children with intermediate level ASD, they are harder to control when conducting the experiments. With proper guidance given by the therapist, they are able to perform through the experiment without much hassle. Given enough time and guidance, children with intimidate level ASD can also benefit from this form of treatment with the virtual pink dolphins.

As for children with severe level ASD, they will less likely benefit from this form of treatment, as most of them have no responses to the instructions given by the therapist. Even when immersed in the 3D immersive room, the therapist has difficulty to guide them to complete a given task.

4.1 Feedbacks of Experiment Devices

The feedbacks are collected from the parents, teachers, therapists, and special education professors who are accompanying the children with ASD during the experiments. Their feedbacks and comments on the virtual pink dolphin serious game are included as follows.

Calibration of Motion Detection

A feedback is about the calibrations of motion detection inputs performed in the virtual pink dolphin serious game. As most of the children with ASD have trouble standing still for certain period of time, the calibrations of the motion detection inputs might be a challenge to them. Only the children with high functioning level ASD have no trouble with the calibrations of the motion detection inputs, as shown in Fig. 10. For the children with intermediate and severe level ASD, they are either having trouble or unable to get the calibrations properly completed.

Sensitivity of Motion Detection

Another feedback is the sensitivity of motion detection devices on tracking the children when they move around. The problem arisen because most of the children with ASD do not stay still long enough in the same spot. Sometimes mid-way through the serious game, some of the children run out of the play area or move toward the screen to try to touch the virtual pink dolphin on the screen. Hence, the tracking of the children, in some cases, will be lost when these situations arise.

3D Glasses

While in the 3D immersive room, children are encouraged to put on the 3D glasses to get the full immersive experience in the virtual pink dolphin serious game. However,

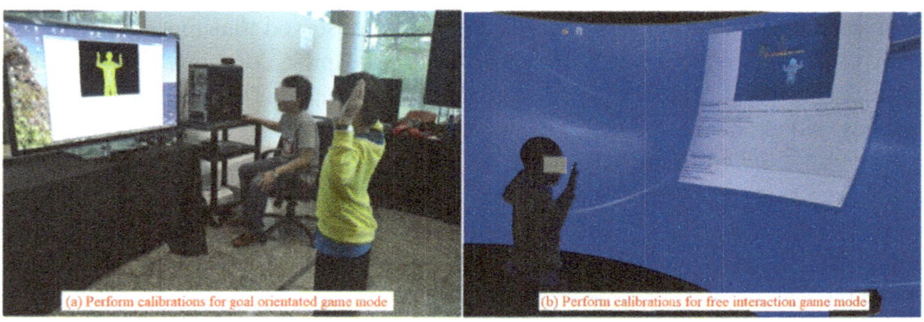
(a) Perform calibrations for goal orientated game mode
(b) Perform calibrations for free interaction game mode

Fig. 10 Children with high functioning level ASD perform calibrations some children are unwilling to put 3D glasses on. The dimension of 3D glasses available in the market is one size for adult participants. The dimension of 3D glasses would not fit the heads of children.

4.2 Feedback on Game Play

Observed from the experiment, special education professors and teachers comment that both the free interaction game mode and goal orientated game mode have too many things going on at one go. The concern is that children with ASD tend to find it hard to concentrate on an object. If the screen is filled with too many different objects, they tend to lose focus on the main objective and concentrate on something else in the background.

For example, in the free interaction game mode, there are three virtual pink dolphins in the lagoon. It is possible for all three virtual dolphins to perform different tasks simultaneously, which might overwhelm the children with ASD. In the free interaction game mode, there are hot air balloons that slowly float up in the background of the lagoon shown in Fig. 11, which is designed to train children with ASD playing the dolphin game with no or little distraction from the balloons.

The music for the free interaction game mode is very soothing to help calm and relax the children with ASD while they play the serious game. It helps to remove some of the anxiety of being in a new environment. It receives positive feedbacks from teachers.

Fig. 11 Example of hot air balloon floating around in the background

4.3 Discussions on Experiment

Through the collaboration with AWWA School, the special needs schools in Singapore are more than willing to choose new technologies to help their students. It is observed that children with ASD are the same as normal functioning kids, in the sense that they also like to use IT gadgets to learn rather than the conventional way. It is also observed that similar methods have been employed to help the children with ASD at AWWA School, e.g., the multi-sensory room, which is quite similar to the immersive room in NTU. It makes easier for the kids to accept the interaction with a virtual pink dolphin in the 3D immersive room.

It is suggested that the virtual pink dolphins can be given more cartoonish faces rather than mimicking the real dolphins. With the cartoonish faces, the virtual pink dolphin in the serious game can show facial expressions that real dolphins cannot do. This is to aid in the teaching of facial expressions and to maintain eye contact, which are two main life skills being taught currently in AWWA School to children with ASD.

It is also suggested that the human avatar in the goal orientated game mode to be added into the free interaction game mode. The human avatar can assist the therapist in the rehabilitation of the children with ASD by having conversions. By having the conversions, the therapist can encourage the children to speak. It will help the children with social awkwardness and train them to have eye contact with the characters on screen while talking to them. Another benefit of having two-way conversions with the virtual characters on screen is that the therapist can use whistle to give commands to the virtual pink dolphins just like the real dolphin trainers at the Sentosa's Underwater World. This can make the experience in the 3D immersive room closer to that of the real DAT.

4.4 Limitations

For the actual DAT, each session can last from 45 min to 1 h within a period of 3 to 4 days. However, for the experiment of the virtual pink dolphin serious game, the duration for each child is only 30 min and only needs 1 day. This is due to the constrain of the availability of the children and those persons involved, i.e., the therapist, and the availability of the equipment, i.e. the 3D immersive room. The result of these constrains are that the experiment could not fully mimic the process of the actual DAT session. Only a small portion of the therapy has been carried out. The results can be improved if the children with ASD are able to sit through a full session. With that result, it is possible to compare children's interactions between the actual DAT and the virtual pink dolphins.

5 Conclusions

In this work, experiments have been conducted with the involvement of 12 children with ASD, professionals, therapists, teachers, special education professors, and parents. The difficulties faced by the children with ASD include the ability to understand instructions for calibrations in the serious game and the ability to concentrate on the serious game procedure. By observing the actions of children in the serious game, deep understanding of the behavior and reactions of the children with ASD are obtained, which are the requirements for enabling the virtual dolphin serious game to be used as a professional tool for the treatment of ASD. The feedback and comments collected in the serious game are useful to improve the design of the virtual pink dolphin serious game.

Besides the children with ASD, some other children from AWWA School with other forms of disabilities have also interacted with the virtual pink dolphin serious game in the 3D immersive room. Their disabilities range from autistic to muscular dystrophy. The children are allowed to try out the serious game with the supervision of the teachers. All of them have positive opinions to the serious game. The future development of this serious game does not have to be constrained to children with ASD, but to other forms of disabilities as well. Children with other disabilities will also be able to benefit from this serious game.

Besides, catering to children with other disabilities, a scaled-down version of the virtual pink dolphin serious game in the 3D immersive room can be provided for more users. The cost of the 3D immersive room in NTU is beyond the budget of most families with autistic children and special needs schools that might want to make use of this technology. Therefore, the cheap version of virtual pink dolphin serious game will highly benefit the more users.

Acknowledgements The authors would like to thank the students, teachers, staffs, principal, and parents of AWWA School for their support, help, and feedback in this research work. Thanks also go to the Institute for Media Innovation at NTU, and Singapore Millennium Foundation for their funding support.

References

1. Cai, Y., Chia, K., Thalmann, D., Kee, K., Zheng, J., Thalmann, N.: Design and development of a virtual dolphinarium for children with autism. IEEE Trans. Neural Syst. Rehabil. Eng. **21**(2), 208–217 (2013). https://doi.org/10.1109/TNSRE.2013.2240700
2. Cai, Y., Chiew, R., Fan, L., Kwek, M.K., Goei, S.L.: The virtual pink dolphins project: an international effort for children with ASD in special needs education. In: Cai, Y., Goei, S., Trooster, W. (eds.) Simulation and Serious Games for Education. Gaming Media and Social Effects. Springer, Berlin (2017). https://doi.org/10.1007/978-981-10-0861-0_1
3. Carlier, S., Van der Paelt, V.S., Ongenae, F., De Backere, F., De Turck F.: Empowering children with ASD and their parents: design of a serious game for anxiety and stress reduction. Sens. **20**(4), 966 (2020). https://doi.org/10.3390/s20040966

4. Checa, D., Bustillo, A.: A review of immersive virtual reality serious games to enhance learning and training. Multimed. Tools Appl. **79**, 5501–5527 (2020). https://doi.org/10.1007/s11042-019-08348-9
5. Feng, Y., Cai, Y.: A gaze tracking system for children with autism spectrum disorders. In: Cai Y., Goei S., Trooster W. (eds) Simulation and Serious Games for Education. Gaming Media and Social Effects. Springer, Berlin (2017). https://doi.org/10.1007/978-981-10-0861-0_10
6. Johnston, D., Egermann, H., Kearney, G.: SoundFields: a virtual reality game designed to address auditory hypersensitivity in individuals with autism spectrum disorder. Appl. Sci. **10**(9), 2996 (2020). https://doi.org/10.3390/app10092996
7. Liu, X., Wu, Q., Zhao, W., Luo, X.: Technology-facilitated diagnosis and treatment of individuals with autism spectrum disorder: an engineering perspective. Appl. Sci. **7**(10), 1051 (2017). https://doi.org/10.3390/app7101051
8. Lu, A., Chan, S., Cai, Y., Huang, L., Nay, Z.T., Goei, S.L.: Learning through VR gaming with virtual pink dolphins for children with ASD. Interact. Learn. Environ. **26**(6), 718–729 (2017). https://doi.org/10.1080/10494820.2017.1399149
9. Medavarapu, S., Marella, L.L., Sangem, A., Kairam, R.: Where is the evidence? a narrative literature review of the treatment modalities for autism spectrum disorders. Cureus **11**(1), e3901 (2019). https://doi.org/10.7759/cureus.3901
10. Miller, H.L., Bugnariu, N.L.: Level of immersion in virtual environments impacts the ability to assess and teach social skills in autism spectrum disorder. Cyberpsychol. Behav. Soc. Netw. **19**(4), 246–256 (2016). https://doi.org/10.1089/cyber.2014.0682
11. Politis, Y., Robb, N., Yakkundi, A., Dillenburger, K., Herbertson, N., Charlesworth, B., Goodman, L.: People with disabilities leading the design of serious games and virtual worlds. Int. J. Ser. Games. 4(2) (2017). https://doi.org/10.17083/ijsg.v4i2.160
12. Radianti, J., Majchrzak, T.A., Fromm, J., Wohlgenannt, I.: A systematic review of immersive virtual reality applications for higher education: design elements, lessons learned, and research agenda. Comput. Educ. (2020). https://doi.org/10.1016/j.compedu.2019.103778

Evaluation of Serious Games for Special Needs Education

Sandra Mei-Yan Chan, Qi Cao, Jieqiong Chen, and Yiyu Cai

Abstract Many serious games have been reported in the literature to help children with special needs on learning and education. Such serious games are embodied with various forms of technologies such as virtual reality, tablets and 3D computer graphics. Different pedagogical theories are also applied in serious games. Evaluation of the effectiveness of serious games varies from qualitative methods to quantitative methods. A virtual pink dolphin (VPD) serious game has been developed in the prior work by the research team at Nanyang Technological University. This chapter describes a preliminary investigation on the effectiveness of VPD serious game for children with special needs. An experiment is conducted with a small number of children from a special needs school in Singapore. They go through two experiment sessions each with three game levels.

Keywords VR serious game · Dolphin-assisted therapy · Evaluation of effectiveness · Game data analysis

1 Introduction

1.1 Background

Serious games have attracted good attention in the recent years for learning, training, and other educational uses. Pedagogical theories and game motivational principles

---

S. M.-Y. Chan · Y. Cai (✉)
School of Mechanical and Aerospace Engineering, Nanyang Technological University, Singapore 639798, Singapore
e-mail: myycai@ntu.edu.sg

Q. Cao
School of Computing Science, University of Glasgow, Singapore Campus, Singapore 567739, Singapore
e-mail: qi.cao@glasgow.ac.uk

J. Chen
VARTEL Network, Singapore 637144, Singapore

Y. Cai and Q. Cao (eds.), *When VR Serious Games Meet Special Needs Education*, Gaming Media and Social Effects, https://doi.org/10.1007/978-981-33-6942-9_7

are integrated into serious games targeting to skill acquisition [11]. There are many serious games embedded with various treatment methods that have been reported in the literature for children with special needs, such as autism spectrum disorder (ASD). Some serious games use virtual reality (VR) and 3D immersive technology to provide better user experiences mimicking the real environment for training children with special needs to perform certain tasks and practise as parts of therapy or treatments. As a kind of occupational therapies, play therapy or role-play therapy is used for children with special needs to express their feelings, thoughts, etc., through game play [9]. Effectiveness evaluation of serious games is subsequently studied to assess the game design with a focus on learning outcome. Evaluating the effectiveness of serious games can be achieved using psychological tests to measure improvements to the individual's abilities after game play, such as memorization ability, mimicry and ability.

Traditionally, the learning progress of children with special needs is assessed by teachers through observation, interview and performance evaluation [10]. Apparently, effectiveness evaluation of serious games is useful for improvements of both serious game design and the learning outcome of students [7].

Data collected in game play can be referred to in-game movement and responses of players, which are acquired by numerical means [6]. Ex situ and in situ methods are typically used in game data collection. Ex situ method records data, which is presented from the "outside" of serious games through observation, together with players' profiles such as age, survey feedbacks, physical pre-test and post-tests. In situ method, on the other hand, collects data presented from the "inside" of serious games, such as timing, scores and number of attempts.

1.2 Objectives and Scopes

The experiment in this research work is conducted to assess the effectiveness of children with special needs to learn knowledge on directions through serious game play with the virtual pink dolphins (VPD) developed in Nanyang Technological University. A virtual avatar is designed to give visual guidance about directions, such as turn right and turn left, using its gestures for children with special needs to follow. Children then mimic these gestures in the game in sequence to give directions to virtual dolphins for them to perform certain actions similar to dolphin shows in many ocean parks.

The objective of this work is to study the effectiveness of VPD serious game for children with special needs in their learning of simple communication skills. The same game may also reinforce other underlying learning objectives such as numeracy, colour and shape. This can be done through the subtle means of the game environment. In the VPD serious game, different colours are being presented in the background graphics such as the orange octopus holding the hula hoop, the blue ocean and the white clouds. Numeracy can also be practised through the number of sets that players are left to complete, while different shapes such as diamonds at each stage are also used as the reward objects, which the dolphin proceeds to catch after each set is completed. The investigation of the VPD serious game has been conducted in two experiment sessions for children with special needs in the ages of 9–18 years old, from a special needs school in Singapore.

1.3 Organizations of the Chapter

After the brief introduction above, the VPD serious game will be illustrated in Sect. 2. This is followed by Sect. 3 depicting on the experiment design and evaluation. Experiment results and analysis are presented in Sect. 4. The conclusions and recommendations are given in Sect. 5.

2 Virtual Pink Dolphin Serious Game

Dolphins are well used in animal-assisted therapy particularly for children with special needs [4, 8]. With modelling, simulation and other 3D VR technologies, the VPD serious game has been developed aiming to substitute real dolphin interactions while maintaining the benefits of dolphin-assisted therapy [3]. This serious game is designed to engage children with special needs for their learning through immersive VR and interactive plays.

2.1 A VR Learning Environment

The VPD serious game is a role-play game. Children with special needs act as dolphin trainers giving instructions to virtual dolphins to perform tricks in the virtual lagoon as shown in Fig. 1. The lagoon is modelled in a fantasy fashion built on top of a turtle.

There are two versions for the VPD serious game developed. A high-end version is designed for 3D immersive room equipped with a 320-degree curved surrounding screen and five sets of active stereographic projectors [2]. The other low-cost version [3] is for single 3D TV display with passive stereographic function. Children wear 3D glasses (active shuttering or passive polarizing) to view and interact with the virtual dolphins through stereographic visualization.

This work is conducted based on the low-cost version with equipments used in the experiment as follows:

(1) Computers with graphics processing unit (GPU) running the VPD serious game software,
(2) 3D Television to display the serious game,

Fig. 1 Role-play VPD game at the dolphin lagoon carried by a turtle

(3) 3D Glasses worn by players, and
(4) Microsoft Kinect devices connecting to the computers for motion sensing purpose.

To better help children with special needs, there is a virtual avatar in the serious game to provide instructions in text, voice and gestures. As players have various sizes on body height and arm length, before the start of the game play, a calibration is performed for the motion sensing devices according to the instructions displayed on the screen. It is to enable the motion sensing devices to detect the players' gestures more accurately.

2.2 The Gesture-Based Gameplay

In the serious game, players are required to conduct five sets of gesture actions mirroring those of the virtual avatar. Each set of gesture actions consists of different sequences of gestures to give directions for game control of the virtual pink dolphins. In each set of gesture actions, players need to execute various combinations of these gestures in sequence according to the prompt of the virtual avatar. These fundamental gestures are listed as follow:

(1) Waving the left hand from left to right continuously,
(2) Forward pointing to the screen with left hand,
(3) Raising of left hand up and down continuously,

Fig. 2 A wrong gesture triggering the "Follow me!" on-screen prompt

(4) Raising of right hand up and down continuously, and
(5) Sweeping from left to right continuously with right hand.

When players correctly execute a set of gestures in sequence, the virtual dolphins perform a trick accordingly. A scoring mechanism is available with the game to record all correct responses of game players. If players execute a wrong gesture, a text "Follow me!" is prompted on the screen with a buzzer sound, to remind players the wrong or inaccurate action, as shown in Fig. 2. When players execute a right gesture, a green tick with an encouragement text of "Excellent!" is displayed on the screen with victory bells ringing. At the end of the game play, players are rated on a 3-star scale for their speed and accuracy, as shown in Fig. 3.

3 Experiment and Evaluation Method

3.1 Experiment Participants

Six students are recruited from the Asian Women's Welfare Association (AWWA) School, a special needs school in Singapore. Consent has been obtained from their parents or guardians of these children before they can participate in the experiment. The participants comprise of those with special needs, including ASD and/or multiple

Fig. 3 Speed and accuracy star rating upon completion of game play

disabilities. Table 1 shows the demographics and characteristics of the students recruited.

Table 1 Demographics and characteristics of the recruited students

Student No.	Profile and characteristics in experiment
1	Autism. Low support needed in experiment
2	Mild intellectual delay and muscular dystrophy. Moderate support needed in experiment
3	Autism. Low support needed in experiment
4	Mild intellectual delay. Low support needed in experiment
5	Autism. Low support needed in experiment
6	Autism. Moderate support needed in experiment

3.2 Experiment Design

Each participant needs to join in two experiment sessions, which are conducted on different dates to play the VPD serious game. In each session, they must complete the entire serious game according to the instructions. Before the experiment starts, there is face-to-face demonstrative guidance by either the teachers of AWWA School or the research team. This is basically to help the participants better understand how to play the serious game at ease. However, some participants may require prompting to complete a gesture during the actual game play in the experiment.

After the completion of two experiment sessions, participants are asked to fill a simple online survey form for the purpose of gathering information including whether they like the game play.

There are three game levels in the experiment as shown below.

(1) Level 1: 1 gesture at each set of action,
(2) Level 2: 2 gestures in sequence at each set of actions, and
(3) Level 3: 3 gestures in sequence at each set of actions.

Participants need to perform five sets of gesture actions in each game level and each experiment session. When the actual experiment sessions begin, participants will be observed, and the number of attempts made before executing the right gesture action is recorded in the forms for further data analysis purpose. The participants are anonymous with no personal information released. Upon the collection and tabulation of the experimental data in Microsoft Excel, bar charts are produced using the pivot table function within the program. The bar charts that are plotted include information such as the number of attempts by each participant for both experiment sessions, to monitor the individual progress and the number of attempts.

Finally, a two-tailed t-test is conducted to evaluate whether the difference between the first and the second experiment session is statistically significant enough. If the difference is significant between two experiment sessions, and the second experiment session achieves better performance, it could indicate that the serious game can benefit the students to learn following and giving directions.

3.3 Statistical Evaluation Method

The t-test is used to evaluate the experiment outcomes when the sample size is small. The t-test is a type of inferential statistics to examine the means of two groups of data sets if there is a significant difference relating to certain features [5]. The t-test calculates statistics according to the two data sets to evaluate whether the assumed null hypothesis, i.e. the means of these two data sets are equal, is accepted or rejected. If the two data sets come from the same population measuring at two different times, a paired t-test is performed. If these two data sets are independent with the same size, the equal variant t-test is performed. Otherwise, the unequal variant t-test is needed.

A two-tailed t-test needs to be performed, to find out whether the means of two data sets are different from one another [1]. A one-tailed t-test is performed, whether to only find out the mean of one data set is greater than or less than the other.

According to the results of t-test, if the p-value of the t-test is less than the alpha level, the null hypothesis is rejected [12]. It means significant differences existing between the means of two data sets. Otherwise, if the p-value of the t-test is greater than the alpha level, the null hypothesis is not rejected.

4 Experiment Results and Analysis

This section begins with the experiment data collection. Following those, the data analysis and discussions are presented next.

4.1 Data Collection

Each participant needs to perform five sets of gesture actions in each game level at each experiment session. For every participant, the number of attempts before the execution of the correct gesture action achieved is manually recorded for the two experiment sessions. The results are tabulated in Microsoft Excel using the pivot table form for quantitative analysis. A t-test is performed to derive the statistic values needed to prove the hypothesis test. This is conducted using the built-in function in Microsoft Excel. The discussions of these results will be elaborated later in this section.

4.2 Data Analysis

This sub-section discusses the data analysis and the outcomes that can be drawn from them. It is done through the analysis of participants' performance at individual game levels. Each participant performs five sets of gestures in sequence at both experiment sessions.

Analysis of Game Level 1

The Game Level 1 consists of one gesture at each set of actions, where participants have to perform five sets of actions at each experiment session. The individual performance of all participants at the Game Level 1 is shown in the bar charts in Fig. 4. The t-test for the data obtained at the Game Level 1 is performed with the results shown in Table 2.

As shown in Fig. 4, there is an obvious decrease in the number of attempts made by the participants in the Experiment Session 2, compared to the Experiment Session

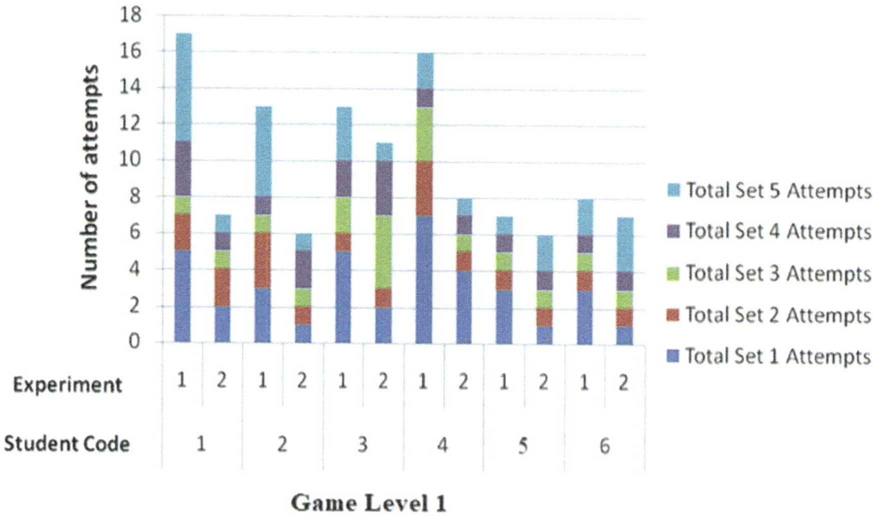

Fig. 4 Individual attempts for Game Level 1 for both experiment sessions

Table 2 Results of t-test performed at Game Level 1

	Experiment session 1	Experiment session 2
Mean	12.667	7.500
Variance	17.867	3.500
Observations	6	6
Hypothesized mean difference	0	
df	5	
t Stat	2.969	
P(T ⇐ t) two-tail (at $\alpha = 0.050$)	0.031	
t Critical two-tail	2.571	

1. According to the results of the t-test shown in Table 2, the mean value of the Experiment Session 2 is lower than that of Session 1 by about 40%. The p-value of the t-test (i.e. $P = 0.031$) is less than the alpha level (i.e. $\alpha = 0.050$). Hence, the null hypothesis is rejected. It means significant differences existing between the means of the data sets recorded in two experiment sessions at the Game Level 1.

Analysis of Game Level 2

The Game Level 2 consists of two gestures in sequence at each set of actions. The participants have to perform five sets of actions at each experiment session. The individual performance of all participants at the Game Level 2 is shown in the bar charts in Fig. 5. The t-test for the data obtained at the Game Level 2 is performed with the results shown in Table 3.

As shown in Fig. 5, there is an obvious decrease in the number of attempts made by the participants in the second experiment session compared to those of the first session. According to the results of the t-test shown in Table 3, the mean value of the Experiment Session 2 is much lower than that of Session 1 by about 56%. The p-value of the t-test (i.e. $P = 0.014$) is less than the alpha level (i.e. $\alpha = 0.050$). Hence, the null hypothesis is rejected. It means that there are significant differences between the means of the data sets recorded in two experiment sessions at the Game Level 2.

Analysis of Game Level 3

The Game Level 3 consists of three gestures in sequence at each set of actions, with five sets of actions per participant at each experiment session. The individual performance of all participants at the Game Level 3 is shown in the bar charts in Fig. 6. The t-test for the data obtained at the Game Level 3 is performed with the results shown in Table 4.

As shown in Fig. 6, there is a decrease in the number of attempts made by the participants in the Experiment Session 2 compared to those of Session 1. But there is an anomaly for the fourth participant, who takes the same total number of attempts before achieving the correct actions at both experiment sessions. Observing the results of the t-test shown in Table 4, the mean value of the Experiment Session 2 is about 30% lower than that of the Experiment Session 1. However, the p-value of the t-test (i.e. $P = 0.053$) is slightly greater than the alpha level (i.e. $\alpha = 0.050$). It means that the

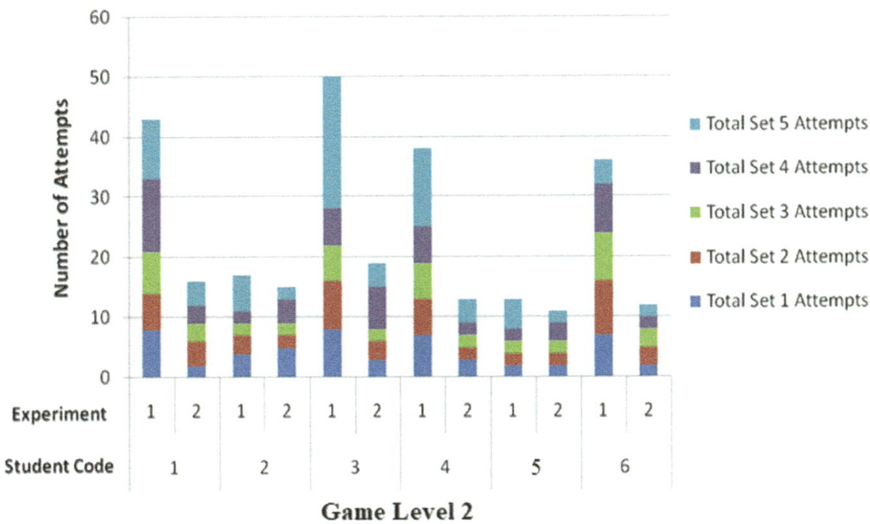

Fig. 5 Individual attempts for Game Level 2 for both experiment sessions

Table 3 Results of t-test performed at Game Level 2

	Experiment session 1	Experiment session 2
Mean	33.000	14.500
Variance	197.200	8.300
Observations	6	6
Hypothesized mean difference	0	
df	5	
t Stat	3.706	
$P(T \Leftarrow t)$ two-tail (at $\alpha = 0.050$)	0.014	
t Critical two-tail	2.571	

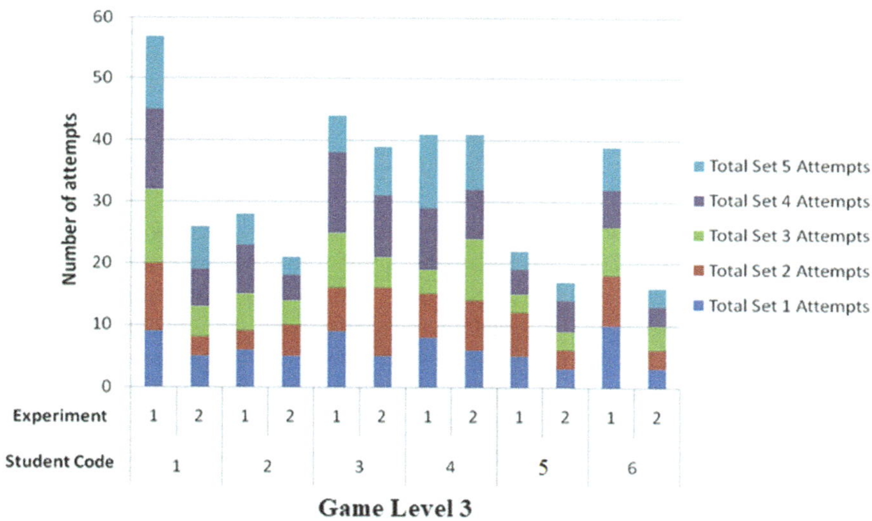

Fig. 6 Individual attempts for Game Level 3 for both experiment sessions

observation between the data of two experiment sessions is not convincing enough to conclude that the average number of attempts differ significantly.

There may be two reasons causing the inconsistency at the Game Level 3, compared to the Game Levels 1 and 2.

- Firstly, it could be the small sample size in the experiment. An anomaly in the small sample size would cause significant impacts to the t-test results. To address this issue, a larger sample size for the experiment is required.
- Secondly, it requires three gestures in sequence per action set, which is more complicated for the participants compared to Game Levels 1 and 2. It is difficult

Table 4 Results of t-test performed at Game Level 3

	Experiment session 1	Experiment session 2
Mean	39.000	26.667
Variance	146.800	119.467
Observations	6	6
Hypothesized mean difference	0	
df	5	
t Stat	2.519	
P(T $\Leftarrow$ t) two-tail (at $\alpha = 0.050$)	0.053	
t Critical two-tail	2.571	

to make improvement for most of the participants in both experiment sessions. An improvement in the scores is limited without significance.

4.3 Survey

The survey is conducted for the participants after they play the second experimental session of serious game. This survey is mainly a means to find out whether they enjoy the VPD serious game. It is not indicative of the effectiveness of the serious game in the learning objectives of following and giving directions. But it is useful in showing the sustainability of the serious game play. While fun and entertainment are not the primary objectives of serious games, it is a fundamentally important motivation factor in the making of serious games for children.

The three survey questions that are asked to students are shown as follows.

(1) Do you like playing the dolphin game?
(a) Yes, (b) No
(2) Which character do you like most?
(a) Dolphin, (b) Octopus, (c) Little man
(3) Which colour in the serious game do you like the most?
(a) Ocean blue, (b) Cloud white, (c) Pink colour of dolphins, (d) Orange colour of octopus

The survey feedback from all participants is shown in Table 5. In the dichotomous style of the first survey question, the response yields a 100% positive. The other two survey questions are asked to find out what parts of the VPD serious game they have enjoyed. These questions reflect the observance skills of the players to the virtual surroundings in the virtual pink dolphin serious game. During the game play itself, numerous students would point out various features they notice in the virtual dolphin lagoon such as the hula hoop, the octopus and the shapes of the gem that the dolphin

Table 5 Survey feedback from participants

Student No	Do you like playing the dolphin game?	Which character do you like most?	Which colour in the serious game do you like the most?
1	(a) Yes	(a) Dolphin	(c) Pink colour of dolphins
2	(a) Yes	(a) Dolphin	(c) Pink colour of dolphins
3	(a) Yes	(a) Dolphin	(c) Pink colour of dolphins
4	(a) Yes	(a) Dolphin	(a) Ocean blue
5	(a) Yes	(a) Dolphin	(a) Ocean blue
6	(a) Yes	(a) Dolphin	(a) Ocean blue

obtained when they are doing the tricks. They also note sound effects such as the countdown timer before each stage begins.

These show that the game is successful in subtly teaching or reinforcing the lessons in shapes and numeracy taught to the children before the serious game. The connection to be drawn from what they have learnt previously and in the game scenes is a good indication of how well children with special needs can relate reality with serious games such as the VPD serious game.

The teachers also give feedback on the effectiveness of the serious game. While the feedback is generally positive, there are areas for improvement. For instance, the virtual avatar can be made bigger with more varied gesture actions.

It is interesting to see during the game experiment sessions that some children wait patiently for their turn and cheer their peers who are playing the game. Encouraging and sharing the joys of accomplishment with each other is an important aspect in the development of children's character.

5 Conclusions and Recommendations

5.1 Contributions

The VPD serious game is effective in helping children with special needs learn to follow directions and then give directions to the virtual dolphins. The improvement in the number of attempts in the experiments shows that the children are better able to follow the directions from the serious game with the second experiment session, although the extent to effectiveness is not significantly large as seen from the t-test analysis conducted in the Game Level 3.

The findings have also shown that serious games are effective in supporting the learning of children with special needs. The suggestions for improvement and for new game concepts to be developed imply that such serious games are indeed useful to teach life skills to children with special needs. VR technology such as the VPD serious game has the potential to simulate in a limitless range of scenarios that can be utilized to teach different skills to the children with special needs. It enables learning happening in various conditions and prepares the children for real-life applications.

From the data of experiment results shown in Sect. 4.2, there is a decreasing tendency with the number of attempts for the second experiment session compared to the first experiment session.

The decrease in the number of attempts by the participants in the second experiment session shows an improvement in their performance. This indicates that they can firstly better follow the directions from the game avatar, and secondly, better imitate and give directions to the virtual dolphins. Thereby, they can execute the correct gesture actions with less number of trials.

From this, it can be implied that the VPD serious game does help to improve the ability of children with special needs to follow and give directions. This is important as children with special needs may not be able to express themselves well in terms of asking for what they need or want. Being able to better give directions helps them in communication and understanding. This skill can be expanded to other areas of life. Being more adaptable to learn to follow directions from the virtual avatar also helps expand the opportunities of learning resources through serious games for children with special needs.

5.2 Recommendations

There are rooms to improve in this study. Firstly, the sample size is on the low end. Secondly, the research will be better if there is a comparison between a control group and an experiment group. A statistical significance t-test can then be conducted to show the differences between the two groups.

Acknowledgements The authors would like to thank the students, teachers, staffs, principal and parents of AWWA School for their support, help and feedback in this research work. Thanks also go to the Institute for Media Innovation at NTU and Singapore Millennium Foundation for their funding support.

References

1. Bevans, R.: An introduction to t-tests (2020). https://www.scribbr.com/statistics/t-test
2. Cai, Y., Chia, K., Thalmann, D., Kee, K., Zheng, J., Thalmann, N.: Design and development of a virtual dolphinarium for children with autism. IEEE Trans. Neural Syst. Rehabil. Eng. **21**(2), 208–217 (2013). https://doi.org/10.1109/TNSRE.2013.2240700

3. Cai, Y., Chiew, R., Nay, Z.T., Indhumathi, C., Huang, L.: Design and development of VR learning environments for children with ASD. Interact. Learn. Environ. **25**(8), 1098–1109 (2017). https://doi.org/10.1080/10494820.2017.1282877
4. Fiksdal, B.L., Houlihan, D., Barnes, A.C.: Dolphin-assisted therapy: claims versus evidence. Autism Res. Treat. (2012). https://doi.org/10.1155/2012/839792
5. Kenton, W.: Tools for fundamental analysis, t-test (2020). https://www.investopedia.com/terms/t/t-test.asp
6. Loh, C.S., Sheng, Y., Ifenthaler, D.: Serious games analytics-methodologies for performance measurement, assessment, and improvement. Springer International, New York (2015)
7. Roediger, H., Karpicke, J.: Test-enhanced learning. Psychol. Sci. **17**(3), 249–255 (2006). https://doi.org/10.1111/j.1467-9280.2006.01693.x
8. Salgueiro, E., Nunes, L., Barros, A., Maroco, J., Salgueiro, A.I., Dos Santos, M.E.: Effects of a dolphin interaction program on children with autism spectrum disorders: an exploratory research. BMC Res. Notes **5**, 199 (2012). https://doi.org/10.1186/1756-0500-5-199
9. Schottelkorb, A.A., Swan, K.L., Ogawa, Y.: Intensive child-centered play therapy for children on the autism spectrum: a pilot study. J. Counsel. Dev. **98**(1), 63–73 (2020). https://doi.org/10.1002/jcad.12300
10. Tahiri, N., El Alami, M.: A new evaluation technique through serious games for children with ASD. Int. J. Emerg. Technol. Learn. (iJET) **15**(11), 202 (2020). https://doi.org/10.3991/ijet.v15i11.12843
11. Tang, J., Falkmer, M., Chen, N., Bölte, S., Girdler, S.: Designing a serious game for youth with ASD: perspectives from end-users and professionals. J. Autism Dev. Disord. **49**(3), 978–995 (2019). https://doi.org/10.1007/s10803-018-3801-9
12. The Pennsylvania State University: Hypothesis testing (2020). https://online.stat.psu.edu/statprogram/reviews/statistical-concepts/hypothesis-testing/p-value-approach

Game-Assisted Vocational Training

Sean Chong, Qi Cao, and Yiyu Cai

Abstract Autism Spectrum Disorder (ASD) refers to variety of developmental disorders that result in challenges in social and communication abilities and repetitive behaviours. While there is no cure for ASD at this moment, it is possible to improve on their area of weaknesses. As of now, therapy is commonly used to improve on the execution skills of children diagnosed with ASD. However, this requires resources, such as time and money, of both the children and the trainers. Game-assisted learning has shown potential to develop the execution skills of children with ASD. This research work covers the design and the usage of serious games to improve children's general execution skills, particularly on basic vocational tasks.

Keywords Executive functioning · Execution skills · Vocational skills training · Serious game learning

1 Introduction

1.1 Background

Executive functioning refers to a set of cognitive control processes including perception and motor responses, enabling self-directed behaviour towards a goal [3]. Skills required in executive functioning comprises of working memory, organizing, planning, inhibitory control, being alert and showing appropriate responses, etc. [4, 9]. Children with autism spectrum disorder (ASD) may suffer from developmental delays in several areas including cognition, social skills, motor skills, etc. [8]. One

S. Chong · Y. Cai (✉)
School of Mechanical and Aerospace Engineering, Nanyang Technological University, Singapore 639798, Singapore
e-mail: myycai@ntu.edu.sg

Q. Cao
School of Computing Science, University of Glasgow, Singapore Campus, Singapore 567739, Singapore
e-mail: qi.cao@glasgow.ac.uk

Y. Cai and Q. Cao (eds.), *When VR Serious Games Meet Special Needs Education*, Gaming Media and Social Effects, https://doi.org/10.1007/978-981-33-6942-9_8

of such deficits by children with ASD is executive functioning [7]. They face challenges with some of these executive functioning skills, such as difficulty on learning or processing new knowledge, difficulty on retaining information in memory, difficulty on multi-tasking, etc. Fortunately, there are methods to help the learning process of children with ASD by teaching processing skills with repeated instructions and practises. Suggested by Bennie [1], parents or teachers can support children with ASD in improving executive function skills by breaking down tasks into smaller and manageable parts, using verbal instructions, using visual supports and visual representation as a reinforcement, and so on.

Serious games are types of educational games to motivate and engage learners in a different way from traditional educational approach [2]. Various formats of information can be integrated in serious games, such as audio, video, voice, text prompts, animations, and interactions. Hamza et al. [5] have presented serious games for vocational training to work in a steel plant and a medical centre. Perini et al. [6] have introduced a serious game for improving the manufacturing skills of young generations and prepare them for the work in industrial world.

Basic vocational skills and housework are easy to learn for most people. But it is challenging for children with ASD to master such basic skills. The challenges associated with ASD result in limited employment opportunities available to those with ASD [10]. Parents and teachers in special needs schools try to teach children with ASD to learn these skills. The learning of vocational skills is in the curriculum of some special need schools. Individualized skills training is necessary to prepare them to join the workforce in the future.

Three-dimensional (3D) serious games are a good form of learning platforms to help children with ASD by providing multiple types of information and prompts for better training outcomes. Tasks can be broken down into smaller steps in game scenes of serious games. Verbal and visual prompts, text instructions, and video animations in serious games can support learning. It is also convenient for children with ASD to re-play 3D serious games for repeatedly practises. Serious games are good candidates to help children with ASD in executive function skills and basic vocational skills training.

It is not a simple task to design an effective serious game to make the game attractive and appealing to children with ASD, as well as easy to operate it. The contents and game flow need be in line with the real-world training, in order to better impart knowledge and skills to children when they play the serious games. The feedbacks and comments from parents and teachers of special needs schools are important to the design of the 3D serious games. The aim of this research work is to design a 3D serious game for vocational skills training of children with ASD, mimics knowledge learnt in the special needs schools, in virtual environment very similar to the real-world scenarios.

1.2 Objectives and Scopes

The objectives are to develop serious games on vocational training for children with ASD. The game-assisted vocational training is to teach them the executive skills, particularly the executive skills required in basic vocational tasks. This work is a collaboration between Nanyang Technological University and METTA School, a special needs school in Singapore. It aims to help train more children with ASD to find jobs in the future. The developed serious game can be operated by most tablets and computers, allowing children with ASD to train their executive skills on vocational tasks at home or at school.

The scopes of this research work include creating 3D models for the game objects using Autodesk Maya, which are then integrated in Unity3D to create the application programs. Device with optical sensor embedded is used to provide direct feedback to the serious game. The combination of the device with optical sensor and Unity3D game engine provides a more realistic gaming experience for them. An analysis of the serious game design is also conducted in this chapter.

1.3 Organizations of the Chapter

The remaining sections of this chapter are organized as follows. Section 2 introduces the ideation and methodology of the vocational skills training serious game for children with ASD. The development and implementation of the serious game are depicted in Sect. 3. The experiment of the serious game and some discussions pertaining to the experiment are presented in Sect. 4. This chapter is concluded in Sect. 5.

2 Ideation and Design Methodology

2.1 Planning Phase

To help children with ASD develop the skill sets on the executive functioning to perform basic vocational tasks, a 3D serious game is proposed to be developed and train children for horizontal surface wiping. It refers to using a cleaning cloth to wipe various dirty horizontal surfaces, such as tables, chairs, bookshelves . This is because horizontal surface wiping is one of the basic skills that children with ASD are required to learn in the curriculum of the special needs school, regardless of the severity of ASD. Horizontal surface wiping is also a skill required in some actual job positions, in which children with ASD might get to apply in the future. Furthermore, horizontal surface wiping is also a skill that is useful when these children return home. They are able to help their family with basic housework that requires this skill, such as wiping the dining table after meals.

As such, a 3D serious game running on computers or tablets can enable the children to practise the wiping motion on common workspace or household horizontal surfaces more conveniently.

2.2 *Tools for Game Development*

Various tools will be used to develop this 3D serious game for vocational training, which are briefly introduced in sub-sections next.

2.2.1 Autodesk Maya

As a 3D computer graphics application, Autodesk Maya, or commonly known as Maya can produce 3D game objects for usage in 3D game development. With Autodesk Maya, game developers can build various desired 3D models and objects from the primitive models. This is done by altering the edges, faces, and vertices of the primitive models. After modelling, Maya allows game developers to generate a UV map of the 3D model. UV mapping is the projection of two-dimensional (2D) images to 3D models. Upon getting UV maps, game developers can design suitable textures for the 3D models. After the modelling and texturing are completed, the 3D models can be exported in a file format suitable to Unity3D software tool for further game development. In this work, the 3D models are exported in the Filmbox file format (.fbx) for Unity3D.

2.2.2 Unity3D

Both 2D and 3D games can be created by Unity3D, a cross-platform real-time game engine. Various graphic APIs such as Direct3D, OpenGL, OpenGL ES, and WebGL are supported by Unity3D. This enables games created to be deployed on different digital platforms. Primitive models can be built from the Unity editor, although it is not easy to create complex 3D models from the Unity editor. In this work, most of 3D models are created in Autodesk Maya or imported from the Unity asset store. In Unity3D, game developers can make use of the primary scripting APIs in C#. It enables more interactions within the game. In this serious game development, programming language C# is used for the scripting API of game objects within game scenes.

2.2.3 Microsoft Visual Studio

Microsoft Visual Studio is a popular Integrated Development Environment (IDE) used to create software programs. It is used in this work to write software scripts for game objects within Unity3D. The scripts are written in C#.

2.2.4 Cleaning Cloth Embedded with Optical Sensor

In this work, a real cleaning cloth is used in the developed serious game. There is an optical sensor embedded in the cleaning cloth. It enables the motions of the cleaning cloth to be detected by the serious game when players move the cleaning cloth for wiping activities.

2.3 *Flow of Game Design*

In this developed vocational training serious game, players can select one of the scenarios to practise flat surface wiping in workspace scenario or home scenario. In the flow of the serious game, there are multiple options of flat surface wiping under each scenario. It is to link to the real-world applications, by simulating various virtual environments.

There are six different virtual workspace environments for players to practise flat surface wiping in the serious game. These workspaces are commonly visited by residents in the real world , which needs to do table cleaning frequently. Some example scenes of virtual workspace environments are shown in Fig. 1. Players can select one of the following virtual workspace environments to practise flat surface wiping.

(1) Virtual food court 1,
(2) Virtual food court 2, with different table colour and setting,
(3) Virtual Starbucks,
(4) Virtual hawker centre 1,
(5) Virtual hawker centre 2, with different table colour and shapes,
(6) Virtual McDonald's, and
(7) Random virtual workspace, which is randomly chosen from the virtual workspaces (1)–(6).

For the home scenario in the serious game, there are six different common household items for players to practise flat surface wiping. Some examples of virtual home environments are shown in Fig. 2. Players can select one of the following household items to practise flat surface wiping.

(1) Virtual dining table,
(2) Virtual tea table,

Fig. 1 Some examples of virtual workspace environments

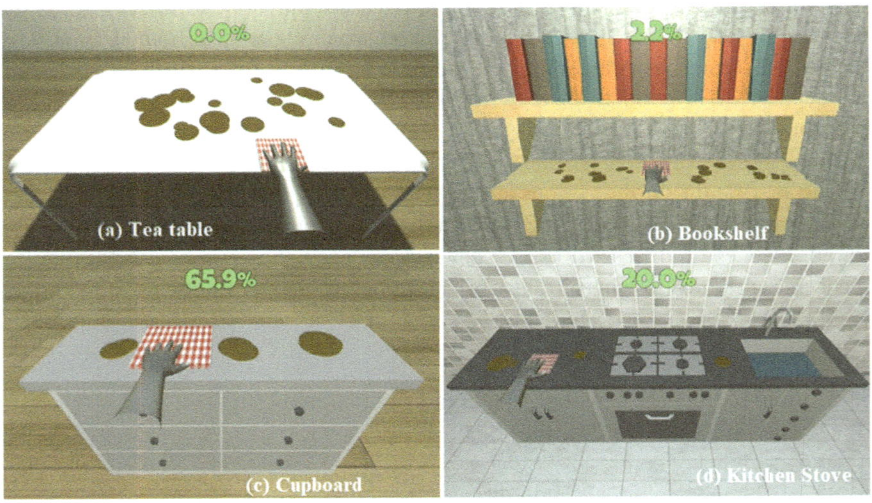

Fig. 2 Some examples of virtual home environments

(3) Virtual bookshelves,
(4) Virtual make-up table,
(5) Virtual cupboard,
(6) Virtual kitchen stove, and
(7) Random virtual household item, which is randomly chosen from the virtual household items (1)–(6).

In the flow of the serious game for vocational training, text instruction prompts will appear to guide players through the actions to be taken. The text instruction prompts appear for about two seconds. After two seconds, the text instruction prompts slide down from the page and disappear. The text instruction prompts help children with ASD to learn what tasks need to do visually.

For each task in the serious game, there is a progress status bar to show progress the percentage of the task. When all dirts and stains are wiped by players, the progress indication will show 100%.

3 Design of the Serious Game

3.1 3D Modelling

In this work, the 3D modelling is done with Autodesk Maya and then imported into Unity3D for the game design. Most of game scenes are designed in Singapore context. The 3D models within game scenarios are designed as close to reality as possible. It is to ensure that children with ASD who are playing the serious game can find these game scenes similar to the real-world environment around their daily life. In this manner, children with ASD can more likely to recognize the vocational skills that they learnt within the serious game are applicable to reality.

When modelling in Maya is conducted, the number of polygons on the 3D models is kept to the minimum. This is to ensure that the exported 3D models will not cause unnecessary performance issues in the serious game. When the number of polygons of 3D models are kept to the minimum, the serious game can run more smoothly without requiring a high-performance computer on 3D rendering. This is crucial for the serious game of vocational training to run on mobile and tablet devices, which cater for more children with ASD.

3.2 Game Scene Design

The game scenes need to be designed individually for each virtual workspace and home scenarios. The game scenes can help children with ASD to learn vocational skills and perform the exercise under different virtual environments.

3.2.1 Randomness of Stain on Surfaces to Be Cleaned

In game scenes, there is a high degree of randomness of wet stains appearing on horizontal surfaces. The rationale of randomizing the wet stains on the cleaning surface is to simulate the randomness of stains in reality. Stains on surfaces are random

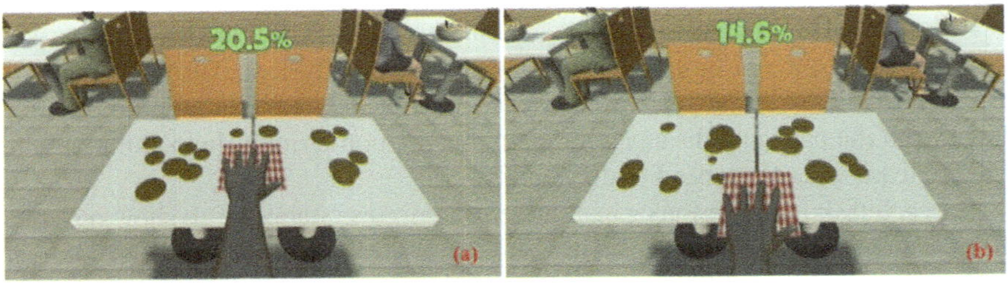

Fig. 3 Random stain appearances at two different instances

and can appear with high variations. For example, stains can appear on surfaces at different positions, different sizes, and different viscosities. Hence, in the developed game scenes, there are three variables of wet stains features that are randomized each time when players enter the game scenes. The three randomized variables are size of stain, position of stain, and strength of the stain. For example, Fig. 3a, b shows the same virtual workspace scenario that are entered at two different instances. It can be observed that positions of the brown stains are randomized between the width and length of the virtual tables. The size of the brown stain is randomized, in which the bigger stains can be up to ten times larger than the smaller stains. The strength of the stains is also randomized, in which more wiping is required to remove the stronger stains. For the stronger stains, each wipe on the stain reduces the size of the stain. This is to simulate stains being wiped in real life.

3.2.2 Variation of Cleaning Surface

In reality, there are countless shapes of tables with different physical features such as size, shape, colour, height, and roughness. In order to allow children with ASD to practise on various horizontal surfaces, there are a total of 13 different cleaning surfaces designed in this game. Each cleaning surface has different physical features such as size, shape, colour, height, or roughness.

3.2.3 Virtual Hand Design

People usually look at positions of their hands when they wipe a cleaning surface in the real world. Hence, a virtual hand, holding a cleaning cloth, has been designed in this serious game. The virtual hand moves according to the motion feedback sent by the optical sensor embedded in a real cleaning cloth. The virtual hand holding onto a cleaning cloth in a table wiping scenario is shown in Fig. 4.

It can also be seen in Fig. 4 that the texture of the hand is not human-like, but metallic. This is a special design consideration for children with ASD according to the feedback of staffs in METTA School. Children with ASD can be frightened when

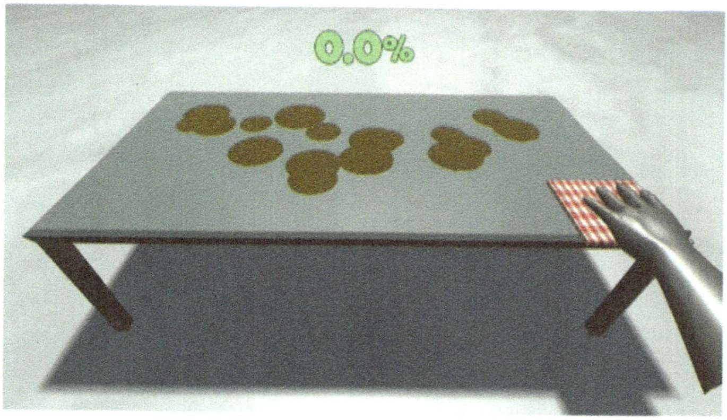

Fig. 4 Starting position of virtual hand on table

they see realistic floating hands in a serious game. Hence, a metallic texture is used instead of a human-like texture.

3.2.4 Initial Cleaning Position

In METTA School, children with ASD are taught to begin wiping from an extreme corner of the table. This eases them in planning the path of wiping of the entire table surface. It will be more difficult for them to plan a path for wiping the entire surface, if they begin wiping from the centre of a table. Hence, to ensure that the serious game for vocational training follows what is taught to children with ASD in METTA School, the position of the virtual hand is initialized at an extreme corner of the cleaning surface in the serious game. Shown in Fig. 4, the starting position of the virtual hand is at an extreme corner of the table. This feature is applied to all cleaning scenarios in the serious game.

3.2.5 Cleaning Progress Indication

Based on previous experiences of teachers at METTA School, some children with ASD are uncertain if they have done a proper job in cleaning a surface. Hence, a cleaning progress indication is implemented in the serious game to aid children with ASD to track their cleaning progress.

As seen from Fig. 5, the progress indication shows how the status of the stain cleaning in each game scene. Children with ASD have to wipe on the stains repeatedly to remove the stains. Upon successful stains removal, the numbers on the progress indication increase accordingly. The task is not completed until the progress indication shows 100%.

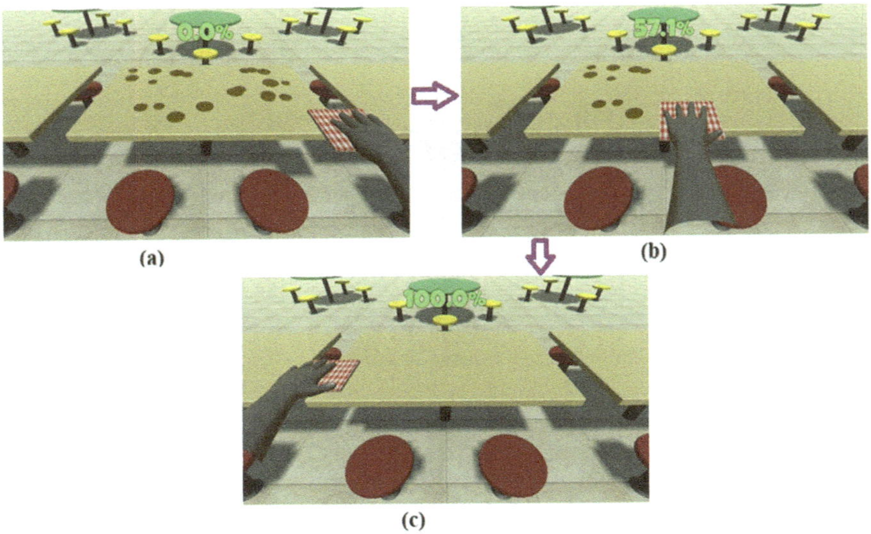

Fig. 5 Progress indication

3.2.6 Visual Prompt

Children with ASD are often guided by visual aids at METTA School. Some children with ASD also require verbal prompts or physical prompts. Hence, upon entering a cleaning scenario in the serious game, there is a visual prompt shown to players for two seconds. It reminds children with ASD that their task is to wipe the cleaning surface.

3.2.7 Hardware Integration with Serious Game

In this research, a real cleaning cloth is used in the serious game. An optical sensor is embedded in the cleaning cloth. This allows the cleaning cloth to send motion feedback to the serious game when players move the cleaning cloth. This provides a more realistic experience to children with ASD, during the gaming process.

The rationale of using an optical sensor is because of its small size relative to a cleaning cloth. This allows simple embedment of the optical sensor to the cleaning cloth. Furthermore, optical sensors can be widely found in computer mouse and are hence relatively low cost.

Table 1 Participants profile and characteristics

No.	ASD profile	Prior assessments on table wiping tasks, by teachers of METTA School
Child A	Higher level functioning	Do not need support
Child B	Intermediate-level ASD	Required partial support from teachers
Child C	Severe level ASD	Required support from teachers

4 Experiment and Discussions

Experiment on the serious game for vocational training has been conducted with children in METTA School. The observations during the experiment are discussed in this section.

4.1 Experiment Setup

Three children with different ASD severity levels have been selected from a class in METTA School. The profiles of these three children are shown in Table 1. Each child performs three experiment sessions through three cleaning game scenes in the serious game. The three cleaning scenarios are food court 1 and Starbucks under workspace environment scenario, and dining table under the home environment scenario.

The equipments used for the experiment include a laptop PC containing the serious game, cleaning cloth with embedded optical sensor, etc.

4.2 Experiment Procedure

For each child, the procedure of the experiment involves the steps as follows.

1. The serious game for vocational training is executed on the laptop PC. The cleaning cloth with embedded optical sensor is passed to the child,
2. On the main menu of the serious game, the food court 1 game scene under virtual workspace scenario is selected for the game play experiment,
3. Under this game scene, the child is instructed to wipe the virtual table with the cleaning cloth, with the cleaning progress being displayed on the PC screen,
4. The child's reaction is observed during the experiment. If the child is unable to wipe the stains on the virtual table entirely, a verbal prompt is provided,
5. If the child still cannot wipe the stains entirely, a physical prompt is provided, and
6. Once the child completes the first game scene successfully, the second and third game scenes are selected next and executed using the same procedure.

Table 2 Experiment results of three participants

Game scenes	Complete without prompt	Required verbal prompt	Required physical prompt
Child A			
Food court 1	No	Yes	No
Starbucks	Yes	No	No
Dining Table	Yes	No	No
Child B			
Food court 1	No	Yes	Yes
Starbucks	No	Yes	No
Dining table	Yes	No	No
Child C			
Food court 1	No	Yes	Yes
Starbucks	No	Yes	Yes
Dining table	Yes	No	No

4.3 Experiment Results

The performances of the experiment of all children are recorded, as shown in Table 2.

It is observed that the Child A who has higher functioning level is able to learn the skills from the serious game quickly, although the verbal prompt is needed in the first game scene for him. After that, Child A is able to independently complete the tasks in the second and third game scenes.

Children B and C need more prompts in the first two game scenes. But, they can catch up the knowledge and complete the third game scene independently. This shows that the serious game can help them improve their ability to locate and wipe all stains off the tables. This is applicable to children with ASD of different severity levels.

It is observed that all children eventually do not require any prompts on the third game scene. This is an important observation, as this would mean that by making use of this serious game, they are able to practise wiping surfaces by themselves without guidance. This could potentially allow children with ASD to practise surface wiping at school or at home by themselves in future.

It should be noted that this is a small-scale experiment that is only conducted on three children with ASD. For a more conclusive result, more experiments need to be conducted on larger sample size of children with ASD.

There is a limitation for the design of the serious game. In the experiment, when the cleaning cloth with optical sensor is moved too quickly, the feedback to the laptop is not fully sent with some data bit losses. This could be because a low-end optical sensor is used for the experiment. In future, a higher end optical sensor can be considered to track the quick motions of the cleaning cloth.

5 Conclusions

A well-designed serious game has the potential to train children with ASD in certain vocational skills. In this work, it has been shown that the serious game for vocational training can help children with ASD to improve basic vocational skills, such as simple table cleaning. The experiment has been conducted by children with ASD from METTA School. It has the potential to allow them to train independently at home or at school in the future. This is done by designing the virtual environment to appear as close to reality and aiming for a realistic gaming experience for the vocational skill. Special game design considerations have been made to accommodate for children with ASD.

As of now, there are already some well-simulated serious games available for children with ASD to play and learn. However, there are not many well-designed serious games to train and improve the vocational skills of children with ASD. With the developed serious game in this work, it can further integrate more functions and develop more game scenarios to train children with ASD in other vocational skills. For example, designing more scenarios allow more variations in the practice of surface wiping. It will be beneficial to the training of children with ASD as they often have problems coping with changes. Besides the training on surface wiping, training on other skills can be added, such as spraying of cleaning solution or water on the cleaning surface.

Acknowledgements The authors would like to thank the students, teachers, staffs, principal, and parents of METTA School for their support, help, and feedback in this research work.

References

1. Bennie, M.: Supporting Executive Function in Children with Autism. Autism Awareness Centre Inc. https://autismawarenesscentre.com/supporting-executive-function-in-children-with-autism-part-2/ (2018). Accessed Aug 2020
2. Cai, Y., Goei, S.L.: (eds.) Simulations, Serious Games and Their Applications. Springer (2013). https://doi.org/10.1007/978-981-4560-32-0.
3. Craig, F., Margari, F., Legrottaglie, A.R., Palumbi, R., de Giambattista, C., Margari, L.: A review of executive function deficits in autism spectrum disorder and attention-deficit/hyperactivity disorder. Neuropsychiat. Dis. Treat. **12**, 1191–1202 (2016). https://doi.org/10.2147/NDT.S104620
4. Gross, R.G., Grossman, M.: Executive resources. Continuum (Minneap Minn) **16**(4), 140–152 (2010). https://doi.org/10.1212/01.CON.0000368266.46038.0e.
5. Hamza, A., Pernelle, P., Amar, B.C., Carron, T.: Serious games for vocational training: a compared approach. In: EEE/ACS 13th International Conference of Computer Systems and Applications (2016). https://doi.org/10.1109/AICCSA.2016.7945712.
6. Perini, S., Luglietti, R., Margoudi, M., Oliveira, M., Taisch, M.: Training advanced skills for sustainable manufacturing: a digital serious game. Procedia Manuf. **11**, 1536–1543 (2017). https://doi.org/10.1016/j.promfg.2017.07.286

7. Shiri, V., Hoseyni, S., Pishyareh, E., Nejati, V., Emami, M., Shiri, E.:Is there any correlation between executive dysfunction and behavioral symptoms in autistic children? A systematic review. Arch. Neurosci. **5**(2) (2018). https://doi.org/10.5812/archneurosci.64303.
8. Tervo, R.C.: Identifying patterns of developmental delays can help diagnose neurodevelopmental disorders. Clin Pediatr (Phila). **45**(6), 509–517 (2006). https://doi.org/10.1177/0009922806290566
9. Turda, E.S., Crisan, C., Albulescu, I.: The development of executive functions among children with autism spectrum disorder. Autism Open Access **9**(4), 243 (2019)
10. Walsh, L., Lydon, S., Healy, O.: Employment and vocational skills among individuals with autism spectrum disorder: predictors, impact, and interventions. Rev. J. Autism Dev. Disord. **1**, 266–275 (2014). https://doi.org/10.1007/s40489-014-0024-7

Design of a Home Bag-Packing Serious Game for Children with ASD

Muhammad Akid Bin Abdul Aziz, Qi Cao, and Yiyu Cai

Abstract The use of the serious games can improve adaptability and planning ability of children with autism spectrum disorder (ASD). Game-assisted learning is gradually more popular in special needs education. In this work, a home bag-packing serious game is designed by collaborating with Institute of Mental Health Child Guidance Clinic (IMHCGC), Singapore. It introduces the methodology and the reasons behind the intentions of implementing certain functions in the serious game to help the development for children diagnosed with ASD. The use of materials are made more accessible and families in order to improve the chances of better adaptation. Game-assisted learning has shown good potential to help children with ASD training at home thus reducing trips to a specialized therapist. The implementation of augmented reality in the home bag-packing serious game is described in this work.

Keywords Adaptability learning · Planning ability training · Parental involvement · Augmented reality serious game

1 Introduction

1.1 Background

A range of conditions associated with autism spectrum disorder (ASD) cause challenges in speech impairment, communication skills, social abilities, cognitive, etc. [6]. The surrounding environments can influence ASD conditions. It results in individuals with ASD the impairment of executive functioning, the inability to learn

M. A. B. A. Aziz · Y. Cai (✉)
School of Mechanical and Aerospace Engineering, Nanyang Technological University, Singapore 639798, Singapore
e-mail: myycai@ntu.edu.sg

Q. Cao
School of Computing Science, University of Glasgow, Singapore Campus, Singapore 567739, Singapore
e-mail: qi.cao@glasgow.ac.uk

Y. Cai and Q. Cao (eds.), *When VR Serious Games Meet Special Needs Education*, Gaming Media and Social Effects, https://doi.org/10.1007/978-981-33-6942-9_9

new skill sets and the difficulty to fit into society or conform to social norms. As mental and cognitive processes to achieve goals in dynamic environment, executive functioning capacities include planning, working memory, impulse control, and inhibition, etc. [3–5].

Therapies have been proposed for children with ASD aiming to slowly overcome some of issues on executive functioning. But many of these therapies need the involvement of trained psychiatrists and training space, which is inconvenient to children with ASD and their families to make the trip down for the treatment. Some serious games have been reported in the literature to provide treatment for children with ASD. Software applications or apps of serious games running on laptop PC, smart mobile devices such as tablets or iPhones are possible solutions to overcome the traditional use of therapy [1]. The components are combined with education, skill enhancement,behavioural changes, and fun learning elements in serious games to heighten interests of those diagnosed with ASD [2]. Such serious games can be played conveniently at home or schools as an educational medium to increase the accessibility for children suffering from ASD. This is beneficial to children with ASD as they can continuously receive the training and education required. Augmented reality (AR) can overlay physical world and environment with digital images and virtual objects [7]. Players can manipulate virtual game objects and mix with real-world surroundings. AR serious games can make this improvement to enhance the education and experience, which would bring benefit to children with ASD.

This work focuses on improving three executive functioning to assist children with ASD, which include planning, adaptability, and inhibitory control functions. The planning function refers to perform a task with regularly changing sequences of actions. The corresponding actions to be executed need to be re-evaluated and updated before they can be executed accordingly. On the contrast of the planning function, it is strictly following procedures and instructions without analysing the changes of sequence of actions. The adaptability refers to the mental flexibility to regulate behaviour by assessing changes in the situation or environment. On the contrast, less adaptability means signs of discomfort or displeasure is caused if there is a change in situation or environment for children with ASD. The inhibitory control means the capacity to regulate behavioural responses, motions or movement. Suffering from inhibitory control issues may display signs of involuntary actions that usually are repetitive such as rocking motions [5].

Improvement can be observed in sociability skills through interacting with virtual characters in AR serious games. Using real environment as the background, such AR serious games can be kept in familiar or comfortable zones of children, so they can more focus on building their socializing ability with virtual game objects through repeatedly practises. It can excite and draw the interest of children with ASD once they get familiar with playing) such AR serious games.

1.2 Objectives and Scopes

This research is a collaboration between Nanyang Technological University (NTU) and the Institute of Mental Health Child Guidance Clinic (IMHCGC), Singapore. The goal is to develop a serious game to help improve the planning, inhibitory control, and adaptability of children with ASD. Aside from the joint development in the research work, IMHCGC also intends to evaluate and monitor the effectiveness of the serious game for the game-assisted learning for children with ASD.

A home bag-packing AR serious game is designed and developed for children with ASD to pack their school bags according to a timetable that is listed for them. There are three virtual rooms in this role-play AR serious game, a bedroom, a kitchen, and a living room. In each room, objects are scattered throughout the room randomly.

As the first objective, players have to transit among three rooms to collect the right items in the stipulated time limit and pack these objects into a school bag before going to schools. Collecting right items into the school bag earns players stars, while collecting wrong objects does not reward any stars. Scoring of playing the AR serious game is based on players' number of switch times from room to room and the number of stars awarded.

After completing the first objective, pack all virtual items in the AR serious game, the next objective is to build a foundation to correlate knowledge learnt in the serious game to objects in real world. In the home bag-packing serious game, the virtual items spawn in an AR are based on image targets in three different rooms. Players have to physically walk to different images that are placed in specified locations in real world. It is different from the first objective by pressing buttons of the serious game to toggle between rooms.

The serious game designed can be used by IMHCGC to help the development of the children diagnosed with ASD under its care. Aside from helping with the development, IMHCGC also intends to monitor and evaluate the effectiveness of using game-assisted learning for children with ASD.

1.3 Organizations of the Chapter

The remaining sections of this chapter are organized as follows. The proposed methodology used in this work is depicted in Sect. 2. Section 3 describes the design details and limitations of the AR home bag-packing serious game. The conclusions and future works are given in Sect. 4.

2 Methodology of Design

2.1 Ideations of Proposed Serious Game

As attempting to help develop executive functioning for inhibitory control, adaptability or mental flexibility, and planning, a serious game is proposed together with the collaborators from IMHCGC. A system to store and process data is expected to be created alongside the serious game. This system can store the performance data of children with ASD who play the proposed serious game. This data must be able to be retrieved and evaluated to compare the improvements in performance of children with ASD along with the game assisted learning.

Besides, parental involvements in the serious game are also expected by IMHCGC. It brings up the idea of a parental component level in the serious game, which can only be accessed by parents. This is to help parents stay aware of the learning progress of their children. It also helps promote more interactions between parents and children with ASD.

The proposed game is not complicated, to reduce the level of difficulty for the children. As such, they can focus better on the learning to improve executive functioning. The lighter mental load can also help the children with ASD not be too pressured thus stay comfortable in the serious game environment. This would allow players to enjoy the serious game.

After brainstorming and ideation, the home bag-packing serious game is proposed where players are required to pack a bag for school according to a daily timetable that changes every day. The objects in the home bag-packing serious game are spawned in different locations in the house. Players are required to toggle among various locations to look for these items.

The proposed serious game for home bag-packing is made up of 10 levels of varying difficulty. Each difficulty level is given a timer of 60 s. Players must complete each level within this time limit. The home bag-packing serious game provides players with three different locations to search for the items needed to pack their bags and prepare for school. The rooms are the bedroom, the kitchen, and the living room. The bedroom is always the starting position of players at each difficulty level.

Players' performances are scored by measuring the number of times of switching between rooms and the number of starts indicating wrong items that have been selected and placed into the bag. The rationale is to assess if players can plan ahead by looking through the timetable first and gathering items that are found in one room before moving on to the next room. If player searches items one by one by following the order in the list, which are scattered among three rooms randomly, it is a clear sign that the player does not analyse the list and plan ahead the next move, to collect all items in the same room first. The serious game also tests children with ASD by having a different list to pack each day to enhance the adaptability. Less points would be awarded for wrong objects that are packed into the bags.

For each level, there is a maximum of four stars to be earned. The stars are not only rewarded players for their performance but also given them a sense of self fulfilment through instant gratification. It can give the serious game a replay value so that players would return in an attempt to achieve more stars. This replay value draws players to play more and help train players repeatedly in the development of their mental functions.

The parental component gives parents the ability to create customized list that is specific to their own children. There are three parental component levels with varying difficulty that affects the number of items to be added in the list. Parents are able to log into the serious game using their credentials. This is to prevent children from accessing the parental component levels to modify the list. Children need to pack their school bags according to the list that is created by their parents. The performance, in this case, would be assessed by the parents themselves. Should children be able to pack all the items into the bag, their parents can sign in into the serious game and tick all the items been successfully collected. It can unlock special stars as bonuses given to players on top of the play of non-parental component levels.

2.2 *Software Tools Used*

Various software tools used to develop this 3D serious game for school bag packing are briefly introduced in this sub-section. The software tool, Blender, is chosen to create 3D models of game objects, before importing into Unity3D to create the serious game with these game objects. The AR serious game is created and finalized in Unity3D, aiming to improve the planning, inhibitory control and adaptability for children with ASD.

2.2.1 Blender

Blender is a free 3D computer graphics modelling software toolset, which can be used to create 3D models. It supports the entire pipeline of 3D development from modelling, simulation, to rendering. It can merge models with different textures giving models with varying looks or effects. A variety of geometric primitives, polygon meshes, and digital sculpting are supported by Blender. Less reliant on dimensional accuracy, Blender relies more on the manipulation of edges or vertices to create various intended models quickly with less memory and higher processing speed. The created objects, meshes, and materials in Blender are organized as forms of data blocks, which can be packed into a single*.blend* file. Blender supports a variety of import and export scripts for 3D file formats interoperable with other 3D modelling tools, such as Unity3D.

2.2.2 Unity3D

Unity3D is a popular game engine framework for game development, which brings together core areas using graphical and physical calculations with various coded functions. Unity3D contains large number of libraries of pre-built 3D models and 2D textures to be invoked for game development. Additional assets or models created by other software such as Autodesk Maya and Blender, can be imported into Unity3D and assembled into game contents. Various types of contents can be edited and optimized in Unity3D. It supports scripts of codes in Java or C# programming languages.

2.2.3 Microsoft Visual Studio

In this work, the gaming scripts are created in C# language, which is compiled and assembled in Microsoft Visual Studio, an Integrated Development Environment (IDE). The scripts of codes with C# can be integrated into Unity3D in the game development.

3 Design of Home Bag-Packing Serious Game

3.1 Game Overview

The home bag-packing serious game is designed as a drag and drop 3D game from the first-person perspective role-play. The screen space of the serious game is optimized to accommodate more furniture in a typical home environment. As such, more realism can be achieved by spawning virtual objects on furniture where it would more commonly exist.

Items are spawned not only according to the furniture that they may be commonly associated with, but also rooms. For example, food is spawned in the kitchen rather than the bedroom. Books are not spawned in the refrigerator. Items' spawn positions are random in a way that objects can spawn in different locations in the same room when the same level is played again. This is to help with the mental flexibility for children with ASD so that they do not get used to repetitive patterns.

The game flow of the home bag-packing serious game starts from clicks on the main menu. There are ten game levels for selection by players. If players are new and have yet to unlock any levels, only the next available level is enabled for selection. The parental component levels are only unlocked after players complete playing a stipulated number of game levels. For example, parental component level 1 is only unlocked for play after players have completed Levels 1, 2, and 3. Parental component level 2 can only be played after Levels 4, 5, 6 have been completed. While parental component level 3 can be played only after the remaining four levels, Levels 7, 8, 9, and 10 have been completed by players.

Once players complete a level of the serious game, some reward stars shown in Fig. 1 are given according to the performance of players. A Help button and a Home button are located at the bottom right-hand corner of the screen. The Help button brings up a help panel that gives instructions to players, should they do not know what to do next. The Home button brings players to the previous menu for them to switch from the home bag-packing serious game to other serious games.

After pressing on the level squares that players intend to play, the level game scenes are loaded for game play. The third level is used as an example to outline the game flow of the home bag-packing serious game. If players select to play the game level 3, an example scenario is shown in Fig. 2. There is a 60 s timer located at the top left-hand corner of the screen. It starts counting down as the serious game progresses. The right side of the game scene shows the calendar and the timetable for

Fig. 1 Game levels for selections

Fig. 2 An example scenario at game Level 3

Fig. 3 View the timetable schedule on what to pack

each day. A text prompt displays for 3 s and informs players the goal in the current level.

Players need to check the timetable for the day to see the list requiring to be packed, before starting to collect the items for packing. This process is shown in Fig. 3a, b. Players need to look at the top right-hand corner of the screen and identify the day. Once the day is known, in this example the day as "Monday", players then need to check the timetable schedule for Monday, by selecting the button with the word "Monday" in it.

The list of items to be packed into the bag is popped up and displayed. Players can then walk around to collect items scattering in these three rooms. The background of the serious game is made by modelling objects such as tables, chairs, and beds in 3D before positioning them in rooms' layout, as shown in Fig. 4.

When an item has been collected, it disappears from the scene of the serious game. To check what item has been stored in the bag, players can click on the "view" button that locates at the bottom right corner of the screen. The pop-up "Packing List" panel appears to display items in the bag. If an object has been collected wrongly, players can remove it from the bag. When this is done, the object disappears from the "Packing List" panel and reappears in the serious game at the location where it originally spawned.

When the packing is completed, players press the "Submit" button to immediately end the current level of the serious game. The "Packing List" panel is displayed with green tick for each correct item collected and red cross for each incorrect or missing item in the bag, as shown in Fig. 5.

Ten seconds after the level ends, a new panel is displayed to show the score. The message "Great Job!" is displayed on the screen if players have performed well, as shown in Fig. 6. The next level of the serious game is also unlocked for players to progress next. If players get less than two stars at a level, the players need to repeat this level until three or more stars are scored.

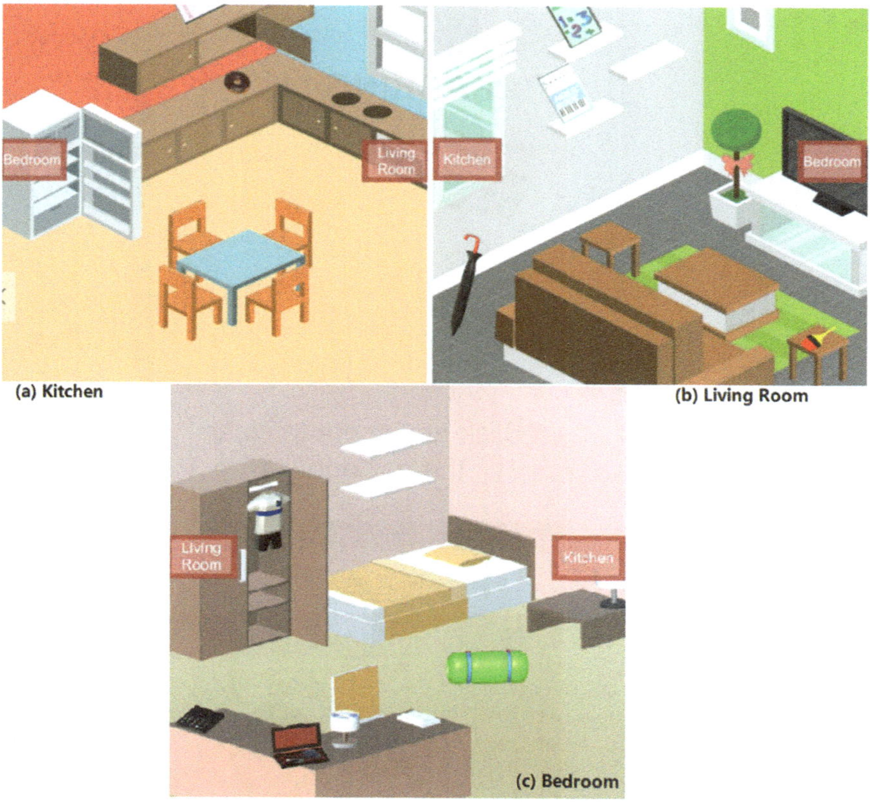

Fig. 4 Rooms environment created for the serious game

3.2 3D Models

The 3D objects are built in Blender. Each object is modelled alongside the real object in an attempt to achieve realism. It takes a minimum of two hours to model a 3D object and another hour for textures. Objects that have many patterns would take longer time to develop textures. This is the same for special textures that mimic various materials from metals to leather.

Besides the pursuit for realism, each object must be easily distinguished for children. For example, pens, crayons, and pencils may share similar shapes. They must be designed with minor exaggerations so that the children with ASD would not confuse themselves. This is similar to the various books designed. Items such as swimwear and school attire have been designed to be gender neutral so that boys or girls with ASD can easily identify them. In order for game objects to be eye catching to children with ASD, game objects are designed in brighter colours than usual. Some examples of created 3D game models available in the serious game such as a pencil case are shown in Fig. 7.

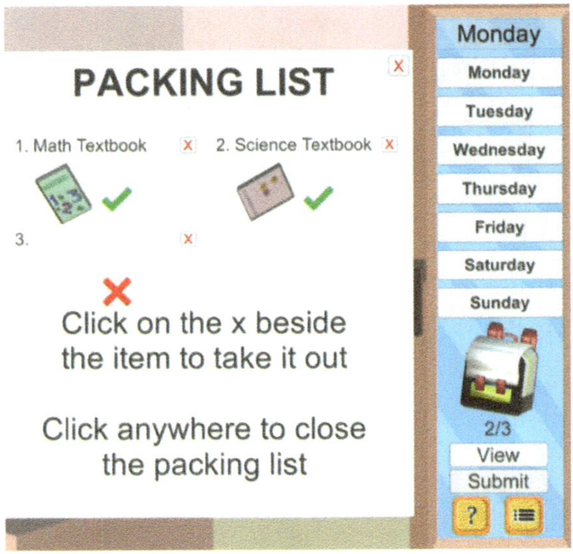

Fig. 5 "Packing List"

Fig. 6 Completed game level

3.3 *Parental Component*

The parental component levels for each sub-games are included in the serious game. The parental component levels are implemented to enhance engagements between parents and children. More importantly, it is to give children with ASD a chance to practise what they have learnt in real life. It aims to train the adaptiveness of children with ASD, for further developing and improving knowledge.

The parents can create a customized list consisting of items in real life that children pack their own real bags with the items in the customized list, as shown in Fig. 8.

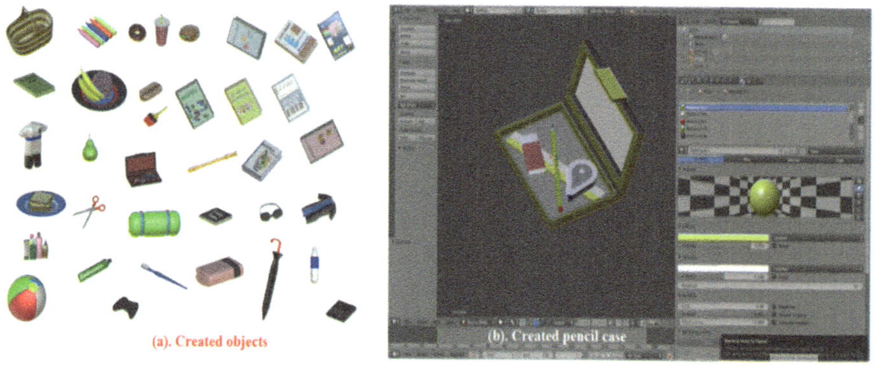

Fig. 7 Various 3D objects been created for the serious game

Fig. 8 Panel to change the list in parental component levels

The parents can then check if their children have accomplished their tasks and mark each item packed with a green tick.

When all items have been packed and verified by the parents, the children will be given in-game bonus stars. It encourages children to strive towards earning more stars and practising their planning, mental flexibility, and inhibition control. Children not only play and get used to the 3D virtual world in the serious game, but also practise in a more dynamic real world. Only parents have access to create the customized list and check the items in the parental component levels. It prevents children from gaining free bonus stars without putting in any effort to play the serious game.

3.4 Scaling of Display Resolution

The serious game is required to support cross-platform compatibility as players may execute the serious game using different devices such as desktop PC, laptop PC, tablet, and mobile devices. The issue encountered in the game development is cross-platform compatibility with the various aspect ratios and scaling according to the screen size and resolution of their personal devices. The solution to this issue is to force the serious game to only operate with the aspect ratio of 4:3 through the setting in Unity3D, as shown in Fig. 9. It would cause devices with wider aspect ratios to experience black vertical bars at the side. But, it is acceptable as the scale of the game objects and texts are still retained. The setting in Unity3D is switched to scale to match the height of the devices' display screens. All texts and game objects are rendered according to the scale.

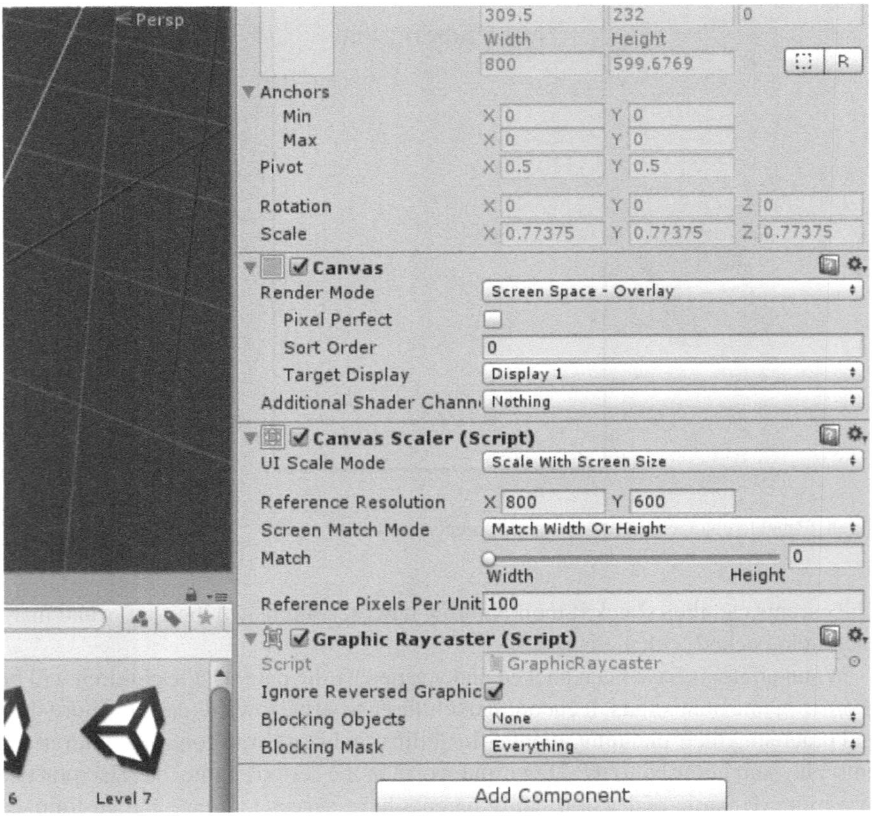

Fig. 9 Panel to change the list in parental component levels

3.5 Augmented Reality

In AR, the field of view of players is determined by moving the camera of the AR device around, which causes game scenes to be constantly changing. This is a challenge to combat the adaptability of children with ASD due to constantly changing virtual environment. In the AR, there are three axes manipulated on data regarding the positions and transformations of all game objects.

When players select a game level, the camera of the AR device is activated. Players need to wander around in the environment to look for the items to be packed. It enables children to use their dynamic surroundings to search items, targeting to train their mental flexibility. When an item is located, players need to point the device's camera to the target. When players select an object, the object can move in 3D space. Initially, the game inputs were coded so that 3D transformation of the game object's position has been done. Players select and drag objects in 3D space to the bag to deposit the game objects.

3.6 Objects Spawning

In the home bag-packing serious game, all objects are spawned randomly in random locations. It means that no two instances of the serious game requires players to pick the same objects to pack into the bag. Hence, it brings better training of the adaptability of children with ASD. This is achieved by giving each game object a number from 0 to 34. Since each object is stored in that array, they can be called out to be instantiated just by using their numbers that are associated with them individually.

Next, the object numbers 0 to 34 are arranged in random order. Certain number of game objects will be selected randomly and spawned in game scenes. The number of objects to be spawned is determined by the game level in which players are playing. If players are playing an easier or lower level, less number of objects are selected thus less objects are spawned.

The location numbers 0 to 5 are also arranged and assigned to selected game objects randomly. It is to place the randomly selected game objects to random spawning locations in the game scene.

3.7 Limitations

In this serious game, the modelling takes too much time due to the detail and volume of 3D models that needed to be done. When interacting with staffs and children in IMHCGC, most of children with ASD lack of information regarding the use of AR. There is a concern of children with ASD confusing and not understanding the virtual game objects projected in an augmented environment versus the physical world around them. It needs to conduct trainings to children before they use the home bag-packing serious game. Otherwise, the use of AR maybe counterproductive by causing children with ASD to not be able to differentiate what is real and what is augmented.

4 Conclusions

A home bag-packing serious game has been designed for children with ASD in this work. The gap of understanding for children with ASD can be bridged through education and learning on the serious game. The home bag-packing serious game has varied number of items to be packed according to the difficulty levels. The items are randomly scattered among three rooms. With every attempt at playing the game, children with ASD are subconsciously practising and training to improve their abilities. It can help them overcome potential problems such as poor planning, poor inhibitory control, and poor mental flexibility. With the AR technology, better serious game can be designed to further train children with ASD to correlate knowledge learnt in the virtual world to the real world.

Future work includes experimental design and data analysis for the serious game developed. Control and experiment comparison can be conducted for children with ASD to see the effectiveness of the serious game designed.

Acknowledgements The authors would like to thank the staffs of IMHCGC, Singapore, for their support, help, and feedback in this research work.

References

1. Cai, Y., Goei, S.L. (eds.): Simulations, Serious Games and Their Applications. Springer (2013). https://doi.org/10.1007/978-981-4560-32-0
2. Carlier, S., der Paelt, S.V., Ongenae, F., Backere, F.D., Turck, F.D.: Empowering children with ASD and their parents: design of a serious game for anxiety and stress reduction. Sensors **20**(4), 966 (2020). https://doi.org/10.3390/s20040966
3. Hill, E.L.: Evaluating the theory of executive dysfunction in autism. Dev. Rev. **24**(2), 189–233 (2004). https://doi.org/10.1016/j.dr.2004.01.001
4. Jurado, M.B., Rosselli, M.: The elusive nature of executive functions: a review of our current understanding. Neuropsychol. Rev. **17**(3), 213–233 (2007). https://doi.org/10.1007/s11065-007-9040-z
5. Kiep, M., Spek, A.: Executive functioning in men and women with an autism spectrum disorder. Autism Res. **10**(5), 940–948 (2016). https://doi.org/10.1002/aur.1721
6. Pisula, E., Ziegart-Sadowska, K.: Social communication and language deficits in parents and siblings of children with ASD—A short review. Autism Spectrum Disorder—Recent Advances. Michael Fitzgerald, IntechOpen. https://doi.org/10.5772/59134
7. Zhu, L., Cao, Q., Cai, Y.: Development of augmented reality serious games with a vibrotactile feedback jacket. Virtual R.ity & Intell. Hardw. **2**(5), 454–470 (2020). https://doi.org/10.1016/j.vrih.2020.05.005

iPad Serious Game to Aid Children with Special Needs in Emotion Learning

Zhi Hao Jeremy Goh, Qi Cao, Jieqiong Chen, and Yiyu Cai

Abstract An iPad serious game in emotion learning is presented in this chapter for children with special needs, such as autism spectrum disorder. Some of the children with special needs suffer in recognizing their own emotional states and/or other people's emotional states. They may lack appropriate strategies to deal with ever-changing emotional states. They have difficulty in knowing how to interpret the emotional states of others and to react to their emotional feelings, such as anger, sadness, and so on. Often their reactions may not match their emotional states. With rapid developments in technology nowadays, mobile devices, including iPad, and their software applications are gaining popularity as an educational tool. It can help create potential treatments for children with special needs. The touch screen and simple human–machine interface make iPad user friendly for children with coordination or learning difficulties. With this technology, it is possible to create serious game applications suitable for children with special needs to learn, understand, and react to emotions in different scenarios. By incorporating embedded videos to mimic basic daily life scenarios, iPad serious games aim to recreate the emotions allowing children with special needs to learn recognizing their own emotional states and the emotional states of others around them.

Keywords Emotion recognition · Emotional expression states · Learning emotions · iPad serious games

Z. H. J. Goh · Y. Cai (✉)
School of Mechanical and Aerospace Engineering, Nanyang Technological University, Singapore 639798, Singapore
e-mail: myycai@ntu.edu.sg

Q. Cao
School of Computing Science, University of Glasgow, Singapore Campus, Singapore 567739, Singapore
e-mail: qi.cao@glasgow.ac.uk

J. Chen
VARTEL Network, Singapore 637144, Singapore

Y. Cai and Q. Cao (eds.), *When VR Serious Games Meet Special Needs Education*, Gaming Media and Social Effects, https://doi.org/10.1007/978-981-33-6942-9_10

1 Introduction

1.1 Background

Researches have shown that facial expressions are related to emotions both biologically and culturally. Communicative values are derived from emotional expressions as outward manifestations of an inner state. Facial movements play a part in influencing one's emotional experience according to the facial feedback. For example, a student is happy because he or she scores good grade in an exam. Usually, it can be perceived how other people feel by judging from their facial expressions, instead of having them to tell their emotional states verbally. With facial expressions, people usually can acknowledge others' actions, express their thoughts and intentions, communicate emotions, etc. However, people suffering from some medical or mental conditions lack awareness of what others are thinking or how they are feeling. Some children with special needs may encounter difficulties in communicating with others using non-verbal cues, leading to failures in building peer relationships [5, 14]. Some of them are incapable of expressing or reciprocating to people emotionally and socially [9]. They have difficulty to understand or predict the emotions of people around themselves. Hence, it is challenging in learning emotions for children with special needs. Autism spectrum disorder (ASD) is a type of special needs with disability causing development delays, social impairments, communication impairments, repetitive behavior, etc. [3, 4]. Children diagnosed with ASD often avoid eye contact [1]. They tend to not play with others, doing activities or tasks alone [6]. The development delays in speaking cause some of them being non-verbal in communication. Visuals or body gestures are more effective ways of communication to them. Constant interventions provide children with special needs learning opportunities to be more sociable and to be more responsive toward social interactions. There are a few intervention methods for children with special needs including behavioral interventions, developmental interventions, and cognitive-behavioral interventions [8]. Behavioral interventions provide structured techniques for tailored behaviors and skills to suit the specific needs of each child. Developmental interventions teach children daily social, emotional, and communication skills to develop positive peer relationships. Cognitive-behavioral interventions shape learning principals and encourage desired behaviors. It helps children to learn how to process information inputs from sights, sounds, and smells. A computer-based multi-sensory serious game is reported to train emotional recognition for children with special needs [16]. Hopkins et al. [11] assess the usefulness of a computer-based social skills training software tool for children with special needs. A computer-based serious game to help children with special needs understand emotions from facial expressions is presented [1]. Tanaka et al. [18] use computer-based intervention to teach face recognition skills to children with special needs. An Internet-based serious game is reported aiming to teach children with special needs to recognize emotions from facial expressions and other cues [9].

Mobile computing devices and tablets such as iPad have increasing computation power with the development of computer technologies. Their touch screen display attracts many users due to user-friendly interfaces. Such devices and software applications provide good platforms as educational tools for children with special needs. Children may feel sliding and tapping on iPad touch screen easier than typing or writing with keyboard and mouse [13]. Usually, there are various learning strengths and weaknesses for children with special needs [15]. For example, auditory and voice information may be challenging to be processed by them [7]. While visual information rendered may be easier to be processed, as children with special needs can be visual learners [19]. Visual supports can better help the teaching and training of basic skills to children with special needs. A serious game, LIFEisGAME prototype-iPad version is reported to enhance emotional understanding skills for children with special needs [2]. A serious game running on mobile devices is introduced to train children with special needs to recognize the emotional expression of a virtual character Tobias [6]. An emotion recognition serious game for assessing social accessibility of individuals with ASD is depicted running on mobile devices [17].

In this research work, Mood Ninja, a serious game using iPad, is developed to train children with special needs in emotion learning, collaborating with Asian Women's Welfare Association (AWWA) School, a special needs school in Singapore. This serious game requires players to input correct answers to the questions on emotions and recognize different emotions level by level in Mood Ninja.

1.2 Objectives

This work aims to develop an iPad serious game to help children with special needs to learn recognizing both their own emotional states and the emotional states of others around them. Educational games have been used by some special needs schools to teach children with special needs daily skills in order for them to be self-dependent. Some special needs schools, such as AWWA School, have utilized IT gadgets to aid in the teaching of children with special needs. The IT gadgets include the iPad, PCs for PC games, and the Xbox Kinect. The development of this iPad serious game can be incorporated into the curriculum of the special needs schools. The iPad serious game can address the unpredictability issue, which is the biggest problem encountered by children with special needs. It can help children with special needs to learn new skills with predictable situations around themselves. In order to make the learning of facial expressions to be more interactive compared to using pictures showing static facial images, the proposed iPad serious game enables showing the motion of the expression, e.g., act of smiling. It can give dynamic quality in the learning. The iPad serious game can include scenarios to help children learn by understanding social situations and re-enact the emotions.

1.3 Scopes

The scopes of this work are as follows:

1. Collect feedbacks from parents and teachers on existing educational games and IT gadgets used in the special needs schools.
2. Collect information on how special needs schools use interactive digital media and technology to help children with special needs.
3. Develop the iPad serious game for emotional learning.
4. Conduct the experiment and get children with special needs to play the developed iPad serious game.
5. Record experiment data and perform data analysis for the experiment.
6. Use analyzed data to suggest enhancement on the iPad serious game for emotion learning of children with special needs.

1.4 Organizations of the Chapter

The remaining parts of this chapter are organized as follows: Sect. 2 presents the hardware and software tools needed for the serious game development. Section 3 depicts the design of the serious game. The experiment, observations, and results are discussed in Sect. 4. The conclusions including contributions and possible future developments for this work are made in Sect. 5.

2 Tools for the Game Development

2.1 Hardware

To develop this iPad serious game, iPad and Mac computer running Macintosh operating system are required. Apple Inc. only allows iOS applications (apps) to be built and distributed with a Macintosh operating system (OS X) on a Mac computer. Apple Inc. offers Xcode, a free development platform, which can only be run on a Mac OS. However, testing or deploying an iOS app onto the iPad or iPhone devices is only possible by purchasing the iOS Developer Program membership fee.

The iPad is a tablet computer designed by Apple Inc., which runs Apple's iOS system. Over the years, there are many different versions of iPad released in the market. The user interface is built around the iPad device's multi-touch screen. The iPad has built-in Wi-Fi and for some models, cellular connectivity. The functions of the iPad are made use to create the app, which would best suit the objectives of the developed serious game.

2.2 *Software Tools*

The iPad serious game is developed using several software tools, including Xcode, GameSalad Creator, Photoshop Creative Suite, and so on.

2.2.1 Xcode

Xcode is an integrated development environment (IDE) toolset for developing software apps of iPhone, iPad, and other Apple devices. It provides user interface design and entire workflow from coding to submitting to the App Store. The Xcode IDE provides app development details, bugs reports, and helps developers on debugging process. Xcode supports source codes by various programming languages such as C, C++, Objective-C, Java, and Python.

2.2.2 GameSalad Creator

GameSalad Creator is a game development platform that allows building 2D games on a graphical user interface (GUI) for describing rules and behavior of game objects without programming codes in C++, Java, and so on. It supports game development in a drag-and-drop fashion with visual editors and a behavior-based logic system. Game objects, named actors, are affected by rules of behavior components. New behaviors can be formed by inserted different rules into existing behaviors such as movement, changing attribute states, and affecting collision.

GameSalad toolbar contains several buttons such as home, scenes, actors, tables, preview, and publish, which allow developers for easy access and navigations in game developments. Under the GameSalad scenes, a sequence of game scenes or flow of game events such as different stages or levels is generated. Each game scene is made up of different actors assigning with varied backgrounds, buttons, and roles to perform the flow of the game.

GameSalad Actors are game objects such as buttons, backgrounds, or overlays that the developers are able to assign few tasks to each game object. GameSalad has a library of tasks in the form of behaviors, images, and sounds that can be assigned to each actor. The combination of these tasks allows each actor to perform a sequence of actions. After assigning the tasks to actors, they are placed in game scenes as the layout of the developed game.

GameSalad Preview Simulator allows developers to check and adjust game settings in the development phase. Preview simulates the actual gameplay as per on an iPad. The preview function enables to address the bugs earlier. It helps to reduce the time spent on testing phase after the development.

2.2.3 Photoshop Creative Suite

Adobe Photoshop Creative Suite (CS) is a software graphics design and image editing tool. It can be used for creating features and editing static images and motion-based contents.

In the Mood Ninja serious game, there are various buttons, icons, and background images with text edited using Photoshop CS.

3 Design of Mood Ninja Game

This iPad serious game, Mood Ninja, aims to create a non-realistic environment to provide the students an escape from reality yet teaching them the required knowledge of emotion recognition.

3.1 Game Flow

Mood Ninja is created in Asian context to better attract special needs students in Singapore, as shown in Fig. 1. The iPad serious game starts off in an Asian setting and uses a storytelling manner to reach out to the students when players click the "Start" button. The story involves a protagonist, Mood Ninja, and his adversary the shadow Dragon, as shown in Fig. 2.

To save the country of Mood and his sidekicks, players have to solve a series of questions. The sidekicks in this iPad serious game for emotion learning are players' fellow school mates who have been chosen as the student models.

Fig. 1 Main menu of Mood Ninja

Fig. 2 Introduction story

This serious game uses a game level system seen in many mobile games nowadays. There are totally 20 game levels in the designed Mood Ninja serious game. In order to proceed to the next game level, players have to answer all questions related to emotions correctly at the current game level. For each question, participants have to press the correct answer twice (similar to double clicks) to lock the answer. It is for players to confirm their intended answers in the iPad serious game. Once the current game level is completed successfully, only then the next game level is unlocked for players to proceed, until all game levels complete. It ensures that players will not skip any questions set. The game flow chart is shown in Fig. 3.

In the experiment of this iPad serious game, players are required to identify the emotions shown in the photos of the student models. There are 11 emotions tested in this serious game, namely: (1) Angry, (2) Ashamed, (3) Confused, (4) Embarrassed, (5) Happy, (6) Scared, (7) Nervous, (8) Proud, (9) Sad, (10) Surprised, and (11) Sorry. In the first few game levels, players are required to choose the correct emotion

Fig. 3 Flow chart of this iPad serious game

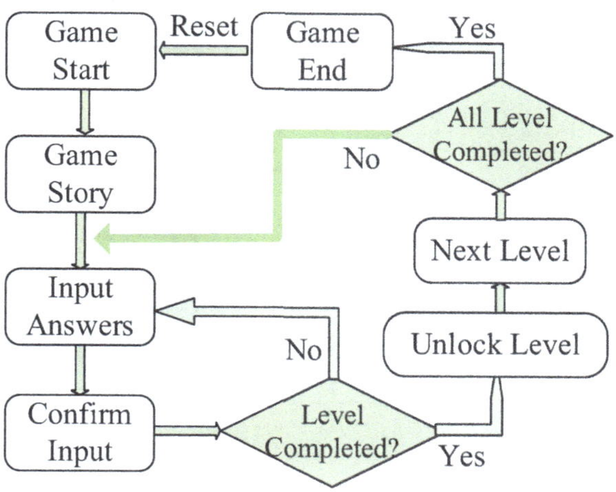

displayed in the photo among four multiple choices of emotional words. In the subsequent game levels, players are required to choose the correct emotion to be displayed in a given scenario by choosing among three photos of the student models displaying different emotions. For every correct answer, a star will be rewarded to players. It will be further emphasized by having correct choice with the star, while having crosses appearing at all wrong choices. At the end of the serious game, teachers can go to the Options menu and tabulate the score of each player.

3.2 Game Implementation

Mood Ninja is implemented using the GameSalad software. The attributes and rules of the GameSalad Actors are programed accordingly to make good drag effects and display effects rendering to iPad devices.

The game scenes are created with size of 1000 × 768 pixels for iPad resolution. The camera size is set to 480 × 320 pixels if converting from iPhone format to iPad, as shown in Fig. 4. A camera actor is created and placed in the middle of the game scene. A drag actor is created and placed in the game scene with one Rule, as shown in Fig. 5.

The menu scene with individual-level actors is created for a level-unlocking system as shown in Fig. 6. An index attribute of *what level am I* (*self.whatlevelami*)

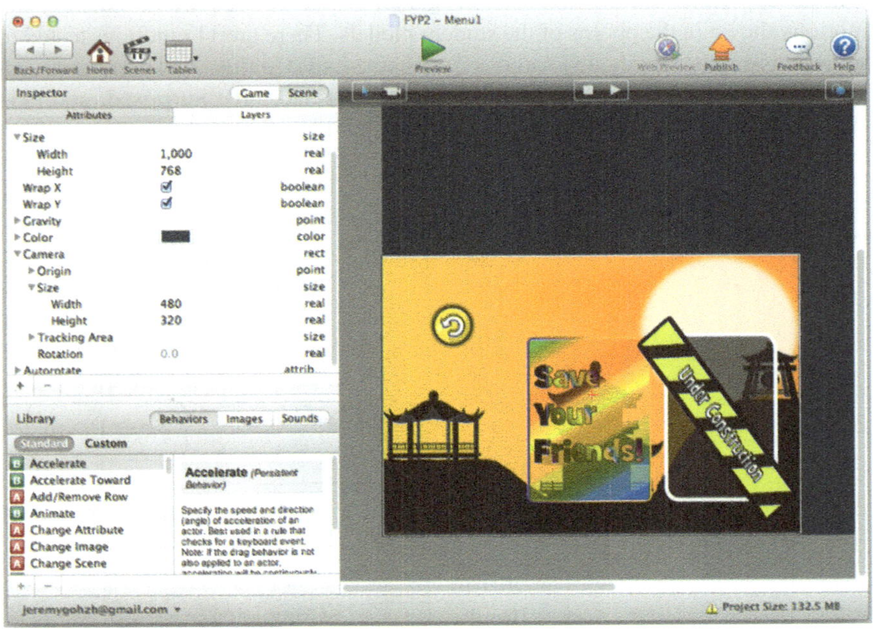

Fig. 4 Setting of scene size and camera size

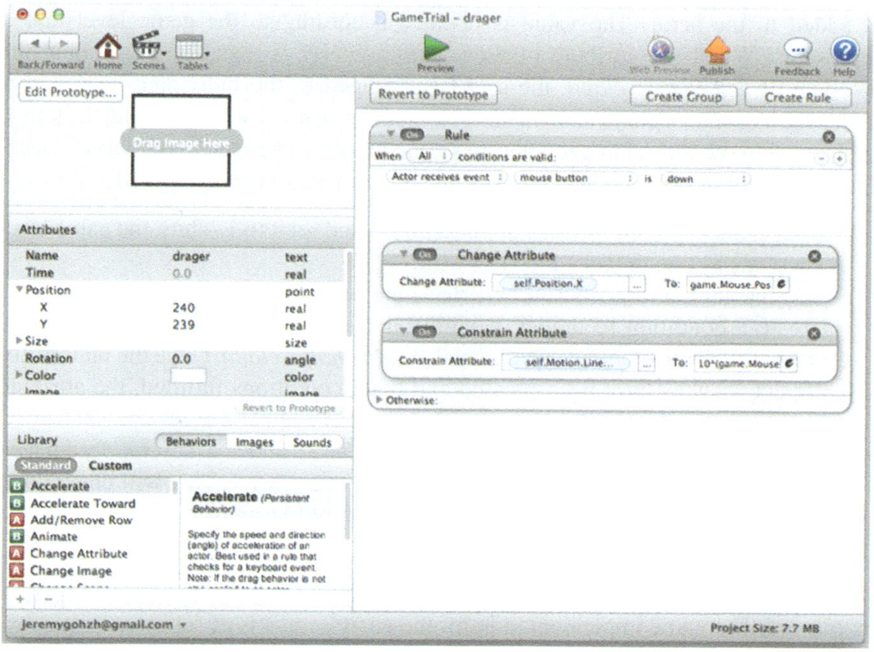

Fig. 5 Drag Actor Rule

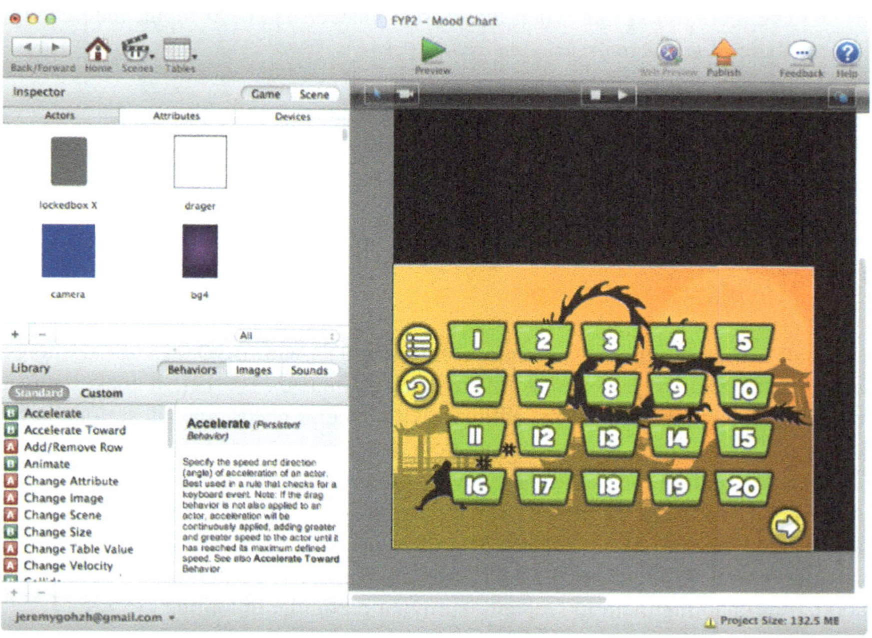

Fig. 6 Game levels unlocking menu

is added to the actor. The value is updated according to the game level that it represents. When the attribute of *game.unlockedlevel* is equal or larger than the attribute of *self.whatlevelami*, the scene is changed to the individual-level scene. When the attribute of *game.unlockedlevel* is smaller than *self.whatlevelami*, the image is changed to the locked image; otherwise, the image is changed to unlocked image. The remaining steps for the level-unlocking implementation are presented as follows:

(1) Add an integer game attribute *unlockedlevel* and set it to 1. Thus, the game level 1 will always be unlocked to start with.
(2) Add integer game attribute *currentlevel* and set it to 1.
(3) Create a condition to unlock it within the individual level. In this instance, change attribute of *game.currentlevel* = *self.whatlevelami*. Once the buttons for correct answers have been pressed and other conditions fulfilled, the attribute *game.unlockedlevel* = *game.currentlevel* + 1.
(4) After that, save the attribute *game.unlockedlevel* under a key *unlockedlevel*. Load game attribute under *unlockedlevel* for every level. This will ensure that the highest value of *unlockedlevel* will be stored so players will not have to unlock from level 1 again after they have unlocked it.

If all game levels have been completed, Mood Ninja displays the scores obtained by players. The game implementation for the game scores is as follows.

(1) Create an integer/real/index attribute and keep it as a counter. In this case, an index attribute of *game.score* is added to the list of game attributes.
(2) Create a condition that when the correct answer has been input, then change attribute of *game.score* = *game.score* + 1.
(3) Under the options menu, create a score panel, as shown in Fig. 7.
(4) Create a score actor and put it beside the star actor.
(5) In score actor, create an attribute called *Display Text*, which calls the value of *game.score*, and display the scores on the score panel.

3.3 Audio Enhancements

Research in the literature shows that most individuals with ASD have good interest and respond positively to audio signals such as music [10, 12]. It makes music an excellent therapeutic tool for children with special needs. In this serious game design, different genres of music are incorporated to suit varied scenarios to attract the attention of children. For example, players are greeted with an upbeat music upon entering the iPad serious game. Happy scenarios are accompanied by upbeat and carefree background music. While sad scenarios are accompanied by slow and moody background music. It creates the ambience and enhances their ability to get into the mood, thus enabling them to feel and learn that particular emotion.

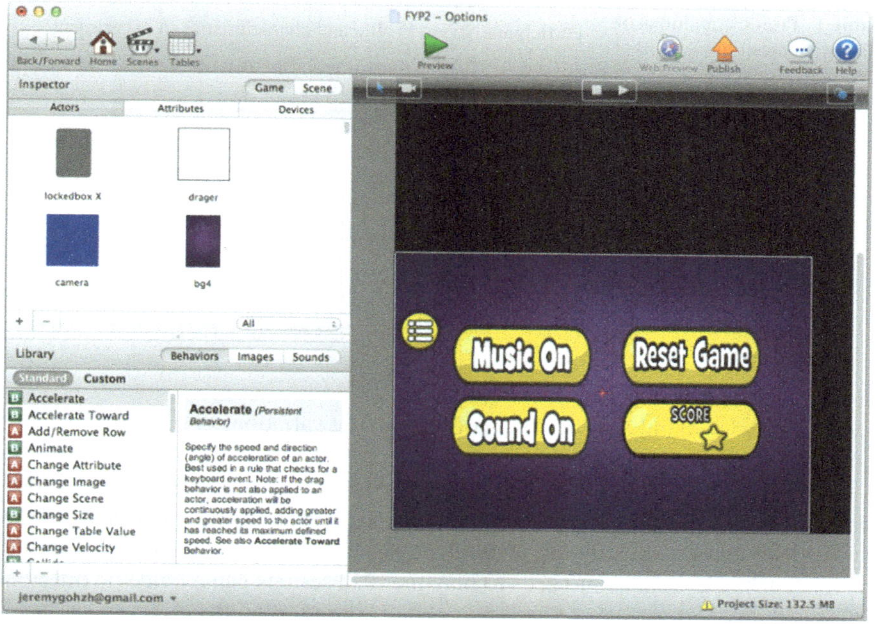

Fig. 7 Score panel under options menu

4 Experiment

4.1 Experiment Procedures

The experimental area is at AWWA School during one of their Information Communication Technology (ICT) classes. This arrangement has minimum interference to their weekly schedule of classes. AWWA School has existing iPad devices for other courses in the curriculum. These iPad devices will be used by participants in this experiment.

The experiment is carried out over 2 days by several psychologists of the school and the research team. Before the start of the experiment, the developed iPad serious game iOS app has been deployed and installed onto six iPads under the school's property. A purchase of a developer license for Apple Developer Program is needed before the serious game could be deployed into the iPads. The location of the experiment is in the IT room of the school, where there are interactive whiteboards, iPads, and computers for the learning under the curriculum.

Seven students from AWWA School are recruited for the experiment, after obtaining consents from their parents and the school. The profiles of these participants are shown in Table 1. There are variations in the profile conditions of the participants with four children with ASD and other participants with lesser cognitive capabilities and multiple disorders. The participants are aged from 8 to 13 years old. Most of

Table 1 Profile conditions of all participants

Participant No	Profile and characteristics in experiment
A	Autism spectrum disorders
B	Autism spectrum disorders
C	Autism spectrum disorders
D	Intellectual disability; Duchenne muscular dystrophy
E	Autism spectrum disorders
F	Premature baby; indication of intellectual disability
G	Global developmental delay

them are male with only one female participant. The variations of participants allow a more neutral test.

At the start of the experiment, a briefing session is first conducted to teach participants how to play Mood Ninja, the iPad serious game. Participants can then start the iPad serious game using their own way with free interaction game play. In the experiment, each participant needs to play the iPad serious game twice on different dates, with the performance of each participant being recorded.

4.2 Experimental Results and Discussion

As some children with special needs have difficulty to understand instructions in the serious game, assistances or prompts by teachers are required to help explain the meaning of these instructions during the experiment sessions. The performances of all participants in the experiment are observed and recorded by the psychologists, with an evaluation form for evaluating their performance (Table 2).

As shown in Table 2, only the first item, i.e., scores, is with quantitative values. The remaining items have not been represented in quantitative formats. In order to better visualize the results comparison differences between the first and second experiment sessions, quantitative values are interpreted for 10 items in Table 2, from *(2) attention* to *(11) App easily understood and responded to.* The cells with *Yes* entry are represented by 1 point. The cells with *No* entry are represented by 0 point, while the cells with *Maybe* entry are represented by 0.5 point. In this case, the performance graph for those 10 items is charted to display the differences in the qualitative format, as shown in Fig. 8. It is observed that the second experiment session exhibits better performance than the first experiment session.

The results will be discussed for each item row by row next.

(1) Scores

The scores of all participants in two experiment sessions are listed in the third row of Table 2. For better visualization purpose, their scores are shown in Fig. 9. The

Table 2 Performances of all participants in two experiment sessions

Participant No.	A		B		C		D		E		F		G	
Experiment session No.	(i)	(ii)	(i)	(ii)	(i)	(ii)	(i)	(ii)	(i)	(ii)	(i)	(ii)	(i)	(ii)
(1) Scores	15	9	24	21	17	10	18	22	2	10	18	18	16	12
(2) Attention	Yes	Yes	Yes	Yes	Maybe	Maybe	Yes	Yes	No	Yes	Yes	Yes	Yes	Yes
(3) Ease of use	Yes	Yes	No	Yes	Yes	Yes	No	Yes	No	Maybe	Maybe	Yes	Yes	Yes
(4) Able to identify emotions from pictures	Yes	No	Yes	Yes	Yes	No	No	Yes	No	No	Yes	Yes	Yes	Yes
(5) Understood emotion words (EW)	No	No	Maybe	Yes	Yes	No	Maybe	Yes	No	No	Yes	Yes	Yes	Yes
(6) Enjoyed the game	Yes	No	Maybe	Yes	Maybe	No	No	Yes	No	No	Yes	Yes	Yes	Yes
(7) Learnt emotions through game	Maybe	No	Yes	Yes	Maybe	Maybe	No	Yes	No	No	Yes	Yes	Yes	Yes
(8) Learnt more vocabulary through game	Maybe	No	Maybe	Yes	Maybe	Maybe	No	Yes	No	No	Yes	Yes	Maybe	Yes
(9) Understood higher difficulty emotions	No	No	Yes	Yes	Yes	No	No	Yes	No	No	Maybe	Yes	Yes	Yes
(10) Able to complete the game	Yes	Yes	Yes	Yes	Yes	Yes	Yes	Yes	No	Yes	Yes	Yes	Yes	Yes
(11) App easily understood and responded to	Yes	Yes	Maybe	Yes	Yes	Yes	No	Yes	No	Maybe	Yes	Yes	Yes	Yes
(12) Better in EW or Picture Scenarios (PS)	EW	EW	PS	PS	EW	Both	Both	PS	NIL	PS	EW	Both	Both	EW
(13) Unable to identify which emotions	3,4,7	2	4,7	7,11	7	2,4,7,10,11	2,4,7,11	10	All except 1,5	2,4,7,8,11	All except 1,5,6,9	3,7,11	11	2,4,7,11

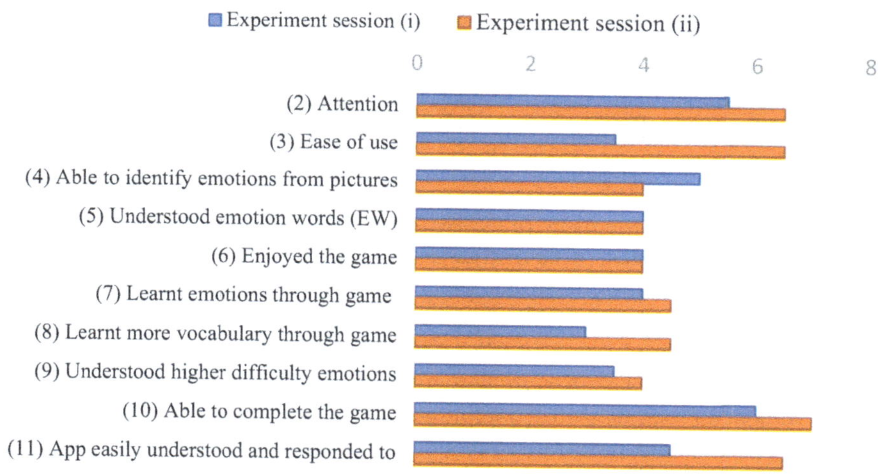

Fig. 8 Quantitative comparisons for two experiment sessions

mean score of all participants in experiment session 1 is 15.7, while the mean score in experiment session 2 is 14.6. There is an overall slight decrease in the mean score in the second experiment session. Participant *E* performed exceptionally poor in the first test. According to the observations and comments of the psychologists, the reason for the discrepancy for participant *E* is that he was not clear what to do in the serious game even after the briefing sessions prior to the experiment. He was having a behavioral issue at that moment. He might be manipulating the situation as well, or he simply does not have the concept of emotions.

As observed from the results, the performances of the participants are not very stable. More experiments need to be conducted in the future in order to identify a trend. In this case, the scores of a participant may not be accurate in identifying

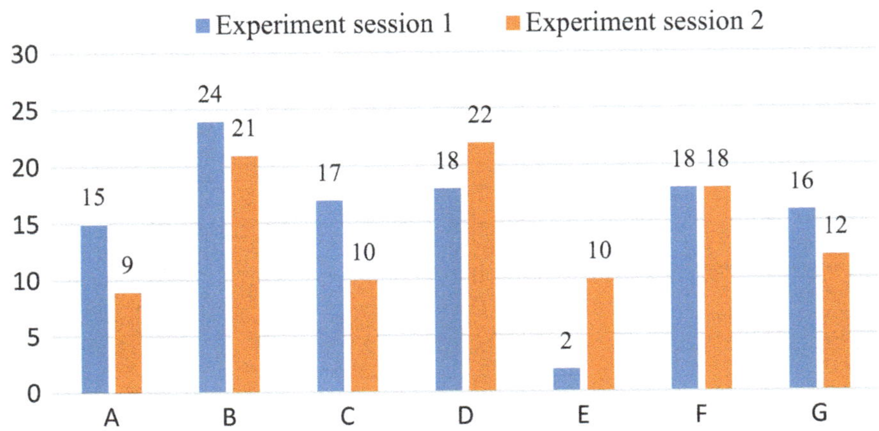

Fig. 9 Bar chart of experiment scores of all participants

the learning progress of the participant. Qualitative methods may be preferred over quantitative methods in this case in evaluating individual's progress. The scores may serve as a benchmark to monitor future progress.

(2) Attention

The attention of all participants in the two experiment sessions is recorded in the third row of Table 2. Most of the participants can focus on the serious game at the beginning. However, they drift in and out of focus throughout the use of the serious game. As observed in the experiment, generally the participants are able to maintain their focus up to the 18th question, while encouragement or inducement of excitement is needed from the psychologists for participants to complete the remaining parts of the experiment. Its reasons are that briefing session is made interesting in narrating the scenarios or explaining the different emotions to attract the attention of participants, but they have short attention span and lose their focus sometime later.

(3) Ease of Use

In the first experiment session, four out of seven observations made are that the iPad serious game is not easy to use for participants without guidance. The reason for this is that some of the features in the serious game are not familiarized when use it for the first time. For example, a safety feature, when inputting the answer, is added to prevent those with muscle disorders from pressing a different answer from what they intended to. Thus, participants have to press the correct answer twice (similar to double clicks) to get it interpreted by the serious game. After explaining and demonstrating the correct method to input answers, participants catch on to it quickly and no further aid is required. In the second experiment session, they find the application much easier to use.

(4) Able to Identify Emotions from Pictures

The observation results for this item are recorded in the sixth row of Table 2. It is observed from the results of both experiment sessions that participants who are not able to identify the different emotions from pictures are also unable to identify the more difficult emotions such as confused and embarrassed. There might be possibility that the virtual models created in this serious game might not display the maximum intensity of an intended emotion. There might also be similarities in the pictures displayed for several related emotions such as sad-sorry and happy-proud. This might affect the accuracy of the score obtained from the experiments.

(5) Understood Emotion Words

Participants who are not able to understand emotion words are also not able to identify higher difficulty emotions. Generally, participants only know four extreme emotions such as happy, sad, angry, and scared. Results that are given by "*Maybe*" is because participants initially do not understand the word. But after explaining the meaning of the emotion word and the context it is used, some participants like *F* can understand and learn the emotion word, while others such as participants *B* and *D* can give fluctuating results. A possible reason could be that participants prefer to stick to their comfort zone or that the pictures shown are correlated in terms of feelings.

(6) Enjoyed the Game

There are four participants giving either a mixed response or not enjoying the serious game at all. This observation made perhaps because the iPad serious game has too many questions. The general feedback from the psychologists is that participants enjoyed the serious game initially. However, toward the end when they lost focus, participants tended to feel agitated. There is not much difference in the results of the two experiment sessions.

(7) Learnt Emotions Through the Game

Participants that are able to identify emotions from pictures correctly initially may or may not learn more about emotions through the serious game. Therefore, in the first experiment, there is a lower result than that of the second experiment. The second experiment has a better result because participants start to perform better in recognizing certain emotions after going through the first experiment session.

(8) Learnt More Vocabulary from the Game

In the first experiment session, there are four observations being "*Maybe*," as it is quite hard to determine whether participants have learnt more vocabulary from the serious game. More observations need to be done to determine whether participants are able to understand and retain the knowledge of the more complex emotions. However, they show the ability to understand the emotions after explaining. In the second experiment session, it further confirms the result that participants learnt more vocabulary from the serious game by repeated playing and prompting by the psychologists.

(9) Understood Higher Difficulty Emotions

In the first experiment session, three out of seven participants show the ability to understand higher difficulty emotions, applying it correctly in corresponding situations. However, another four participants are unable or uncertain in understanding higher difficulty emotions. In the second experiment session, there is a slight improvement in their ability to recognize higher difficulty emotions. Further learning components might be needed in the serious game to reinforce and educate the emotions.

(10) Able to Complete the Serious Game

In the first experiment session, all participants are able to complete the serious game except for participant *E*, being with behavioral issues at that moment. In second experiment session, all participants including participant *E* are able to complete the serious game.

(11) App Easily Understood and Responded to

In the first experiment session, four out of seven participants show clear indication that the iPad serious game app is easy to understand and use. In the second experiment session, six out of seven participants give indication that the serious game app is easy to understand with only 1 "*Maybe*." The improvement in the results shows that the features of the serious game have received a good response from the participants.

(12) Better in Emotion Words (EW) or Picture Scenarios (PS)

Participants give a mixed response but generally, they are better in EW levels than PS levels. This means that participants understand the concept of the emotions but may not be able to identify it correctly in real life. The possible reasons include that they may not be able to identify these emotions directly, or the quality of the emotion displayed in pictures of the serious game needs to be enhanced. In both experiment sessions, a few participants maintain one stronger area. There are two participants improving to be strong in both EW and PS levels, while another two participants decrease in their results slightly to be strong in only one area.

(13) Unable to Identify Which Emotions

In total number of 20 game levels, there are 11 different emotions for learning. In both experiment sessions, most participants are not able to identify emotions like Confused, Embarrassed, Nervous, and Sorry. This is expected because these are not basic emotions. An educational package by the teachers or psychologists may be needed to further reinforce the teaching on higher difficulty emotions.

4.3 Feedbacks of Experiment

There are some feedbacks given by the psychologist and teachers of AWWA School who are present during the experiment of the serious game for emotion learning.

4.3.1 Buttons

It would be better if the buttons are designed bigger and more obviously, or visual indicator being added to signal participants, which buttons can be pressed in order to proceed to next steps.

4.3.2 Game Level Layout

Some participants are unable to recognize that they are on a new level after completing a level. As the layout for every game level looks identical, they have difficulty in differentiating the different levels. In the future development of the serious game, there should be some indications to remind participants that they are on a new level.

4.3.3 Game Play

The game design is good in a way that it acts as a teaching tool for the ongoing Social & Emotional Learning (SEL) class of AWWA School. The SEL class uses role-play approach for children, in order to put themselves in different real scenarios, to understand the underlying emotions and feelings. This iPad serious game allows similar interactions by exposing and re-enacting more virtual scenarios in a safe environment. More importantly, children with special needs are able to learn from it.

5 Conclusions

5.1 Contributions

An iPad serious game, Mood Ninja, is developed in this work, with experiment being conducted at AWWA school. It aims to assist teachers and psychologists to educate children with special needs the emotional learning through game play. From the feedback of the psychologists assessing participants of the serious game, it is shown that several participants enjoy and benefit from it. The benefits include learning new emotion vocabulary to accurately express how they are feeling. They learn the context of the emotions of which they might be feeling in a particular scenario. Due to time constraint, the research work is unable to investigate the long-term effects of emotional learning through this serious game.

This developed serious game can be applied to other special needs schools and also mainstream schools in teaching emotions through game play. By changing the pictures of the student models, the serious game can also be made for public to benefit more people. As iPhones are readily available in the market and have a large amount of users, the serious game can also be resized to iPhone format to play it on both iPhones and iPads.

5.2 Recommendations

Several key recommendations are identified in order to improve the next phase of development. Since the attention span of children with special needs is very short, the amount of questions can be lessened or more game scenes need to be created in between the questions to motivate children to continue playing the serious game. Some possible game scenes to be added including matching faces displaying the emotions. The second recommendation is that more animations, narrations, or sounds are added to capture the attention of children with special needs.

Another possible recommendation might be to research the effect of colors and emotions and to apply in the serious game. A palette of colors can be used to "check-in" the emotions of participants, which will then display the corrective actions for participants to perform. This is because children with special needs may not be able to identify their own emotions that they are experiencing.

5.3 Future Developments

Future development of this serious game for emotion learning includes having videos embedded into the serious game, customizable scenarios by teachers, etc., to make the serious game learning more effective.

From this experiment, feedbacks from teachers and psychologist suggest the serious game to be designed for both iPad and iPhone devices. In today's world, cell phone is seen as a necessity, but iPads are more of luxury. As iPads may be too expensive for some families, iPhone is a more common platform. It makes the serious game more accessible and acceptable to the masses to allow emotional learning to be taught on-the-go. In the case of AWWA School, the school already has a number of iPad devices that are shared throughout the cohort. It makes the iPad serious game acceptable and affordable as there is no need to purchase new iPad devices.

It is suggested that the pictures for the selections on the game GUI can be animated to enhance the expressiveness rather than static images. By having dynamic images on the screen will make the serious game more interactive and more personal.

Acknowledgements The authors would like to thank the students, teachers, staffs, principal, and parents of AWWA School for their support, help, and feedback in this research work.

References

1. Abirached, B., Zhang, Y., Park, J.H.: Understanding user needs for serious games for teaching children with autism spectrum disorders emotions. In: World Conference on Educational Media and Technology (2012).
2. Alves, S., Marques, A., Queirós, C., Orvalho, V.: LifeisGame prototype: a serious game about emotions for children with autism spectrum disorders. PsychNol. J. **11**(3), 191–211 (2013)
3. American Psychiatric Association (2013). Diagnostic and Statistical Manual of Mental Disorders, 5th edn. https://doi.org/10.1176/appi.books.9780890425596
4. Barendse, E.M., Hendriks, M.P.H., Thoonen, G., Aldenkamp, A.P., Kessels, R.P.C.: Social behaviour and social cognition in high-functioning adolescents with autism spectrum disorder (ASD): two sides of the same coin? Cogn. Process. **19**, 545–555 (2018). https://doi.org/10.1007/s10339-018-0866-5
5. Baron-Cohen, S., Golan, O., Ashwin, E.: Can emotion recognition be taught to children with autism spectrum conditions? Philos. Trans. R. Soc. B Biol. Sci. **364**(1535), 3567–3574 (2009). https://doi.org/10.1098/rstb.2009.0191
6. Carvalho, V., Brandão, J., Cunha, P., Vasconcelos, J., Soares, F.: Tobias in the zoo—A serious game for children with autism spectrum disorders. Int. J. Advan. Corp. Learn. **8**(3), 23 (2015). https://doi.org/10.3991/ijac.v8i3.4897
7. Cohen, M.J., Sloan, D.L.: Visual Supports for People with Autism: A Guide for Parents and Professionals. Woodbine Hpuse, Bethesda (2007)
8. Corsello, C.M.: Early intervention in autism. Infants Young Child. **18**(2), 74–85 (2005)
9. Fridenson-Hayo, S., Berggren, S., et al.: Emotiplay': a serious game for learning about emotions in children with autism: results of a cross-cultural evaluation. Eur. Child Adolesc. Psychiat. **26**(7), 979–992 (2017). https://doi.org/10.1007/s00787-017-0968-0
10. Gold, C., Wigram, T., Elefant, C.: Music therapy for autistic spectrum disorder. Cochr. Database System. Rev. (2006). https://doi.org/10.1002/14651858.CD004381.pub2
11. Hopkins, I.M., Gower, M.W., Perez, T.A., et al.: Avatar assistant: improving social skills in students with an ASD through a computer-based intervention. J. Autism Dev. Disord. **41**(11), 1543–1555 (2011). https://doi.org/10.1007/s10803-011-1179-z
12. Kaplan, R.S., Steele, A.L.: An analysis of music therapy program goals and outcomes for clients with diagnoses on the autism spectrum. J. Music Ther. **42**(1), 2–19 (2005). https://doi.org/10.1093/jmt/42.1.2
13. KarenTBTEN: The iPad: a useful tool for Autism. HubPages. https://hubpages.com/health/ipad-for-autism (2014)
14. Erik, M., Schuller, B., et al.: The ASC-inclusion perceptual serious gaming platform for autistic children. IEEE Trans. Games **11**(4), 328–339 (2018). https://doi.org/10.1109/TG.2018.2864640
15. Randi, J., Newman, T., Grigorenko, E.L.: Teaching children with autism to read for meaning: challenges and possibilities. J. Autism Dev. Disord. **40**(7), 890–902 (2010). https://doi.org/10.1007/s10803-010-0938-6
16. Serret, S., Hun, S., Iakimova, G., et al.: Facing the challenge of teaching emotions to individuals with low- and high-functioning autism using a new serious game: a pilot study. Mol. Autism **5**, 37 (2014). https://doi.org/10.1186/2040-2392-5-37
17. Sturm, D., Peppe, E., Ploog, B.: eMot-iCan: design of an assessment game for emotion recognition in players with autism. In: IEEE International Conference on SeGAH (2016). https://doi.org/10.1109/SeGAH.2016.7586228
18. Tanaka, J.W., Wolf, J.M., et al.: Using computerized games to teach face recognition skills to children with autism spectrum disorder: the Lets Face It! Program. J. Child Psychol. Psychiat. **51**(8), 944–952 (2010). https://doi.org/10.1111/j.1469-7610.2010.02258.x
19. Tissot, C., Evans, R.: Visual teaching strategies for children with autism. Early Child Dev. Care **173**(4), 425–433 (2003). https://doi.org/10.1080/0300443032000079104

Design of a VR Supermarket Serious Game

Jun Hong Goh, Qi Cao, and Yiyu Cai

Abstract Virtual reality (VR) is used as an aid for developing intervention skills for children with autism spectrum disorder (ASD). The advancement in technology has enabled the capability of VR to be more readily applied for special needs education. A virtual supermarket serious game is created to help children with ASD to develop independent life skills . Eventually, it helps them adopt the ability of shopping on their own. An experiment has been conducted with children from a special needs school in Singapore to evaluate the effectiveness of the developed serious game. Participants in the experiment use gesture commands to control their movements in the virtual supermarket. The experiment result shows that participants have improvement over a period of 2 weeks in two experiment sessions. It indicates that VR with motion sensing is a feasible tool for developing new skills for children with ASD.

Keywords Life skills training · Shopping skills · Motion sensing · Virtual supermarket

1 Introduction

1.1 Background

Autism spectrum disorder (ASD) includes diagnoses of autism, Asperger's disorder and pervasive developmental disorder [1]. The cause of ASD has not been fully discovered yet [3]. Research has been going on for a long time, but no direct cure being found. Some findings relate the cause of ASD to family genetics. Children diagnosed

J. H. Goh · Y. Cai (✉)
School of Mechanical and Aerospace Engineering, Nanyang Technological University, Singapore 639798, Singapore
e-mail: myycai@ntu.edu.sg

Q. Cao
School of Computing Science, University of Glasgow, Singapore Campus, Singapore 567739, Singapore
e-mail: qi.cao@glasgow.ac.uk

Y. Cai and Q. Cao (eds.), *When VR Serious Games Meet Special Needs Education*, Gaming Media and Social Effects, https://doi.org/10.1007/978-981-33-6942-9_11

with ASD tend to have social difficulties [6]. Many of them have difficulty in handling their daily life situations [8]. Several aids and therapies have been developed to help children with ASD. Over the years, researchers have been proving the use of serious games as one of the aids in education and social intervention development for children with ASD. Compared to traditional learning approaches, children tend to express more interest towards learning approach using virtual reality (VR) serious game. It motivates children to learn in a safe and controlled environment [7]. Sharing the same objectives as traditional learning aids, VR can help children with ASD train social communication skills, daily life skills and independent skillsets, before practising in real life [11]. It allows them to stay focus on particular tasks, thus benefitting them ultimately. The integrations of VR and augmented reality (AR) technologies into serious games have been implemented by many prior works in literature [9, 10]. Players have various forms of control tools as human–machine interface in virtual environments of serious games. Some examples of such control tools include keyboard, mouse, joystick or headgear [7], which are beneficial for children to train their fine motor skills. But the gained virtual experience has certain difference from reality. Technology advancement on motion sensing and tracking devices enables players to interact in virtual environment more naturally, reducing the gap between virtual and real experiences. Motion tracking devices allow players to use body motions or hand gestures to represent their desired actions in virtual environments. Thus, it improves user learning experiences in VR serious games . It does not require any gear to be worn on players, making it less distractive and more user-friendly for children with ASD.

Daily activities like shopping and purchasing necessities are essential for most people. However, it turns out to be a difficult task for children with special needs. They usually take longer time than others to familiarise with new environments, such as shopping malls and supermarkets. For example, a VR serious game reported to help people with Down's syndrome learn shopping in a supermarket [2]. An in-game small virtual shop is designed for individuals with ASD to understand the concept of currency or money in a VR game [4]. Children with ASD may feel uncomfortable with many unfamiliar people around them in such shopping places. Hence, VR serious game can provide a platform for training children with ASD to get familiar with various environments virtually, before they experience in the physical world. The advantage of VR serious game is the ability to set specified constraint to virtual environments and allow customised development to cater for different virtual environments. It accommodates the needs of every individual with ASD of different level. New features or objects can be added into virtual environment of the serious game according to different learning objectives.

In this work, a developed virtual supermarket serious game aims to create a conducive environment for children with ASD to learn the procedure of making a purchase in the supermarket independently.

1.2 Objectives and Scopes

In this research, a virtual supermarket serious game with motion tracking and interactive capabilities has been developed. It is used to train children with ASD how to do shopping in the virtual supermarket according to the provided shopping list. Although shopping in supermarkets may be easy for most of the ordinary children, it can be a challenging task for children with ASD in real life. With the virtual supermarket serious game, children with ASD can learn and practise shopping skills in a safe manner under controlled environment with the guidance of teachers or parents. They can learn knowledge such as recognising various types of products in virtual supermarket, correlating text information in shopping list to objects in supermarket and looking for items from different categories at corresponding sections in supermarket. Such education in the serious game can train children with ASD independent life skills.

A fully functional virtual serious game involves many aspects from modelling to game implementation. The design details of this supermarket serious game will be presented in this work. An experiment with children from a special needs school, Asian Women's Welfare Association (AWWA) School in Singapore, will be conducted. The experiment scopes include activities as follow:

(1) To familiarise children with ASD in the supermarket serious game,
(2) To develop shopping skills for children in the virtual environment, and
(3) To develop independence for children with ASD in the reality by adopting the skills, knowledge and experience in the virtual environment.

1.3 Organisations of the Chapter

The remaining parts of this chapter are organised as follows: Sect. 2 presents rationales and concepts of the virtual supermarket serious game to be developed. The objects' modelling for the serious game is introduced in Sect. 3. Section 4 describes the design and implementation of the virtual supermarket serious game. Sect. 5 discusses the experiment conducted and its results. This chapter is concluded in Sect. 6.

2 Concept of Virtual Supermarket Serious Game

2.1 Conceptualisation of Virtual Supermarket

The general idea of the virtual supermarket serious game allows children with ASD to move around in the supermarket, shop for the products and make a purchase at the cashier counter. However, to attract the attention of children with ASD is always a challenge. One way to overcome this is to include items realistic enough in the virtual supermarket, which gets them interested. As such, the Power Card Strategy [5] is adopted. The full wide range of products available in the supermarket might not be necessary, as it might cause distraction to children with ASD. In fact, the variety of products should be narrowed down to the products, which most interest to children.

2.1.1 Different Difficulty Levels

Progressive learning is one approach for developing a skill for children with ASD. Thus, the virtual supermarket serious game can be broken down into different levels of difficulties. It aims to facilitate and guide children with ASD slowly on skills development.

For a start at the easy game level, the supermarket serious game is introduced without the appearance of other people, to reduce the uncomfortableness of children with ASD. It means that players are alone in the virtual supermarket. There is a designated route to prompt players to sections in the supermarket that the desired items to be purchased are located. At the shelves, players can choose what they want to buy freely, without a shopping list. Once the product is selected, it is automatically placed into a virtual trolley. Upon completion, players are directed to the cashier counter through an animation demo.

Progressively in the next game level, players are able to navigate around the supermarket freely. Subsequently, virtual shoppers will be added into the supermarket to create a more realistic environment. The selection of products requires players to move them from the shelves to the virtual trolley. In addition, players can go to different sections in the supermarket to continue shopping for products of different categories. Background sound is added into the virtual environment to simulate the crowd noises in the supermarket. The noise level is varied according to the size of the crowd in the virtual supermarket.

The final game level introduces another two features: first, the introduction of returning an undesired product to the shelves; second, the payment for the products at the cashier counter. The payment process introduces the money concept into the virtual supermarket game, so as to complete the entire shopping experience. However, players first need to understand the concept of money, before proceeding to the final game level.

2.1.2 Game Flows

In Singapore context, there are many neighbourhood supermarkets located at residential areas, surrounding by HDB buildings, public housing in Singapore. The supermarket serious game starts with an introductory session which brings players across the road, to the entrance of the opposite neighbourhood supermarket as shown in Fig. 1. At the supermarket entrance, players have the chance to choose shopping items from different categories including drinks, snacks and toiletries. Through

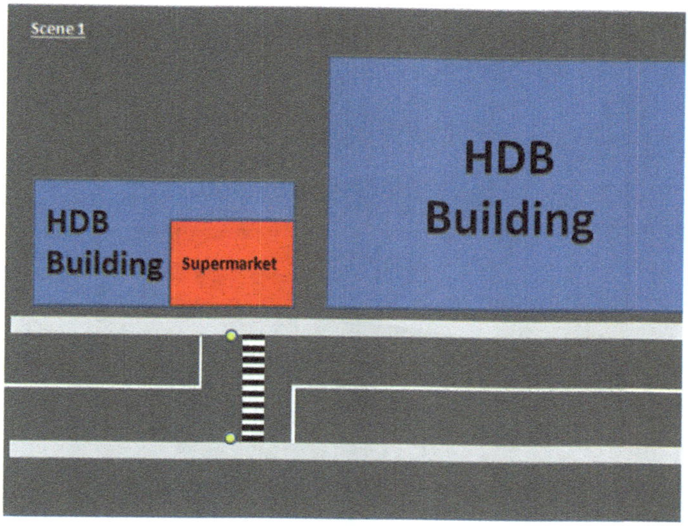

Fig. 1 Top view of the virtual environment in the serious game

another animation demonstration, it brings players to locations where these items are. Upon reaching the locations, players are brought to a game scene, facing a wide range of items under the same category. These items are placed on shelves where players get to choose the desired items and place into a virtual trolley. Players need to decide what items and how many items they want to get from the shelves. To complete the shopping, players need to select the "Cashier" button in the supermarket serious game. The serious game ends with an animation showing players walking towards the cashier counter.

2.2 *Research on Supermarkets in Singapore*

Supermarkets are a very common sight in Singapore. They are located at shopping malls, neighbourhood areas and even at some petrol stations. There are a few supermarket big players in Singapore, namely Fair Price, Giant Hypermarket, Prime Supermarket and Sheng Siong Supermarket.

Fair Price is the largest supermarket company in Singapore, with 230 retail outlets. In general, these supermarkets sell items ranging from food, drinks, daily necessities, household appliances, etc. They have similar layout and format that vary in sizes, with a typical one as shown in Fig. 2. The items are usually classified under different categories and the items range from different brands and sizes. Labels and signages are located at various places in the supermarket. Price tags are clearly indicated on the shelves for the reference of the shoppers. However, the significant difference

Fig. 2 A typical layout of a Fair Price supermarket

is the location of each category of items in each supermarket, due to the floor plan limitations of various outlets.

The supermarket serious game focuses on a more generalised layout. The layout is catered for children to get familiar with the environment in the virtual world. Items in the virtual supermarket are narrowed down to the needs and interests of children with ASD. The idea is to attract children's attention and concentrate better in the virtual environment during the gameplay, using popular items among children. It also includes items, which children come across daily and are more familiar with. Moreover, some are taken reference from the canteen of AWWA School, the special needs school, as shown in Fig. 3.

3 Objects Modelling of This Serious Game

3.1 Modelling of the Serious Game

Models of the supermarket serious game consist of three main sections. They are terrain, supermarket layout and the products. In every section, it involves three stages: research, modelling and mapping.

Fig. 3 Popular selling items in the canteen of AWWA School

At the research stage, it comprises of data collection on the range of products available, the facilities used and the environment layout of the supermarket. Internet research is done to source the products' dimensions and images. The products' appearances are important to enhance user experience in the serious game and help children with ASD to better relate to reality.

At the modelling stage, it involves careful planning of the required details of each game model. These models are created with Autodesk 3ds MAX, a three-dimensional (3D) modelling software. Different modelling approaches result in different polygon sizes. More polygons allow to illustrate more complexity of the model. As shown in Fig. 4, smoothness of the curvature by adding more polygons to the model is compared. Considering the total number of models to be created in the supermarket serious game, the number of polygons contained in each model needs to be carefully calculated. This ensures the serious game to run smoothly without any overloading of polygons in general. Overloading of polygons in the serious game may cause lag and distraction to players.

Models can be converted into "editable poly", which enables various modifications to the shape of models. Five types of selections are available for the modification in

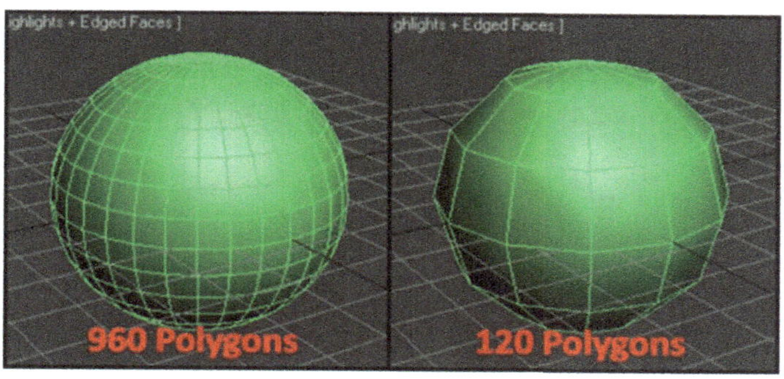

Fig. 4 Model smoothness differs by the number of polygons

Fig. 5 Created models before and after material mapping

3ds MAX tool: Vertex, Edge, Border, Polygon and Element. In the development of the supermarket serious game, three out of five selections are commonly used for shape modelling, i.e. Edge, Polygon and Vertex.

At the mapping stage, material mapping is added onto the surface of the models to create realistic appearances in virtual world. This mapping method is also known as texturing. Figure 5 shows a few game models completed with material mapping used in the supermarket serious game. Material mapping enhances the game models and improves the quality of the serious game.

3.2 *Product Model Methodology*

There are many different approaches to achieve the desired shapes in modelling. For each object, the game model is created with the following steps.

The models created in Autodesk 3ds MAX for the entire supermarket serious game are shown in Fig. 6. These game models are built according to the actual appearances and dimensions of each item. A close-up view of some created product models is shown in Fig. 7. The main advantage of modelling by Autodesk 3ds MAX

Fig. 6 Range of products in virtual supermarket

Fig. 7 A close-up view of some products

is the capability to create models with any forms and shapes. In addition, the scales of models can be easily adjusted to match the proportions in the entire terrain.

In order to illustrate steps of how to create game models for this supermarket serious game, one example of creating the game model is demonstrated next. A sequence of steps describe the modelling methodology for a bottle drink, which is commonly seen in supermarkets.

(1) Start Autodesk 3ds MAX tool. Click "Files" and import the image of the bottle drink taken by a camera. Under Views on the toolbar, go to viewport background. Check "Match Bitmap" under Aspect Ratio and check "Lock Zoom/Pan".
(2) Click on "Create" icon and then "Shape" icon. Select "Line" and draw lines tracing half of the bottle as shown in Fig. 8. The tracing does not need to be too accurate.
(3) At Modifier panel, select "Vertex" to reposition each vertex to match with the image. Be sure to select the "Select and Move" icon. Then, under Selection and Geometry rollout at the modifier, activate the "Fillet" button.
(4) Select one of the points that need to be rounded. Drag the point until it matches with the image at the background. Repeat this step on points that are required to be rounded, as shown in Fig. 9.
(5) At the Modifier panel, select "Lathe". Then click "Y" direction and "Min" button. The lathe is performed in direction of the axis Y.
(6) Convert the model into an editable polygon. Add UVW Mapping at Modifier panel. Select polygon at editable poly rollout and highlight the polygon where the materials are needed to be mapped (Fig. 10).
(7) Press M key for the Material Editor to appear in the screen. Choose an empty slot and insert the mapping. Click the blank box beside diffuse and double click

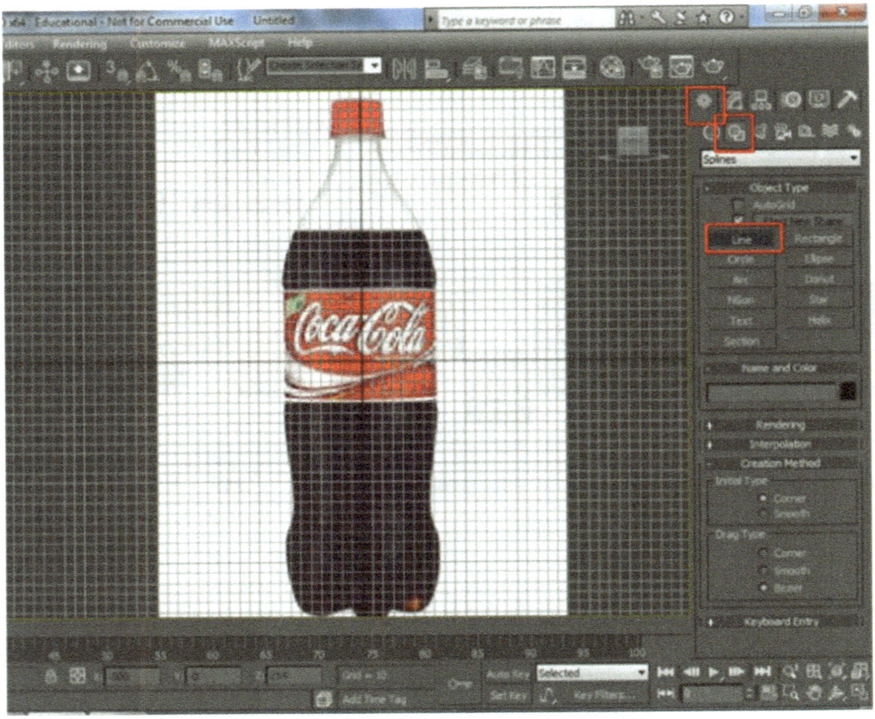

Fig. 8 Create line in 3ds MAX

on bitmap. Select the image from the file folder. Then, drag the material from Material Editor to the highlighted polygons on the model, as shown in Fig. 11.

(8) At UVW Mapping, change the U Tile to two units. This makes the image double in the direction of U. Check "X" on Alignment and click the "Fit" button.

(9) Repeat Step 7 till the completion of the remaining mapping of the model (Fig. 12). Note, only images require resizing need to use UVW mapping, otherwise use normal mapping.

All other game models in this serious game are created following the similar procedures mentioned above. All game models created in Autodesk 3ds MAX are saved in *FBX* file format, which can be imported into Unity3D game engine later.

3.3 Virtual Supermarket Layout and Terrain

The virtual supermarket includes design of facilities and the design layout within the supermarket. The facilities in the supermarket consist of shelves, refrigerators, freezers, trolleys and cashier counters. Figure 13 shows the view of the virtual super-

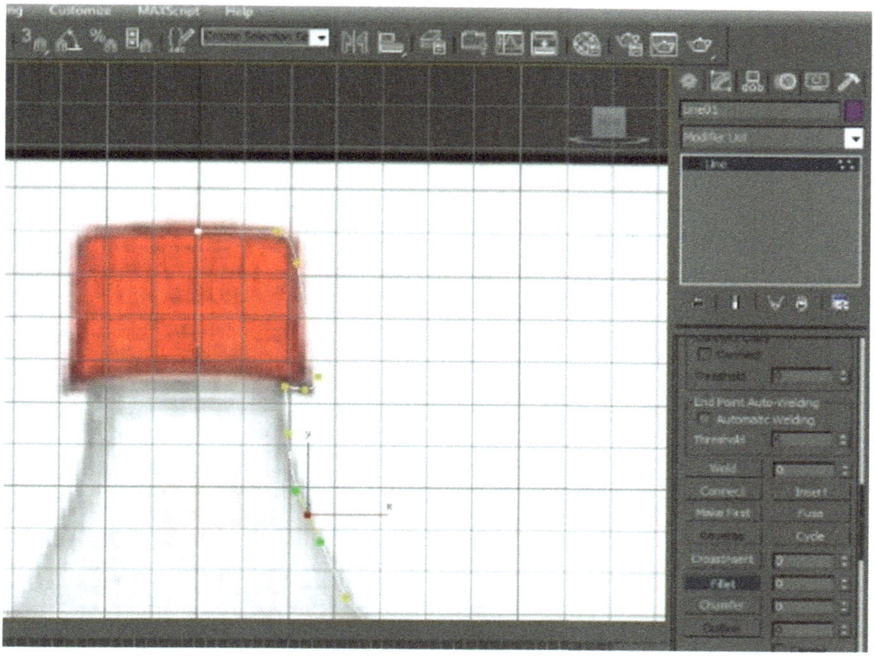

Fig. 9 Modifier Fillet at 3ds MAX

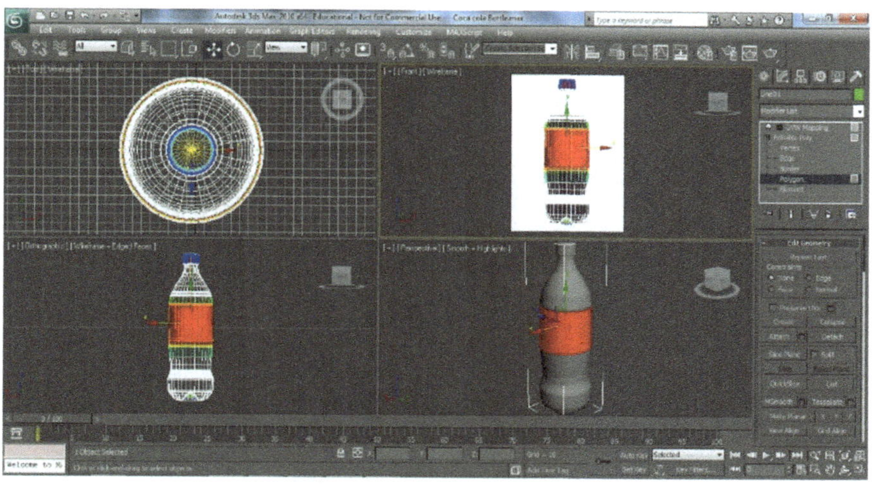

Fig. 10 Modifier panel add UVW map

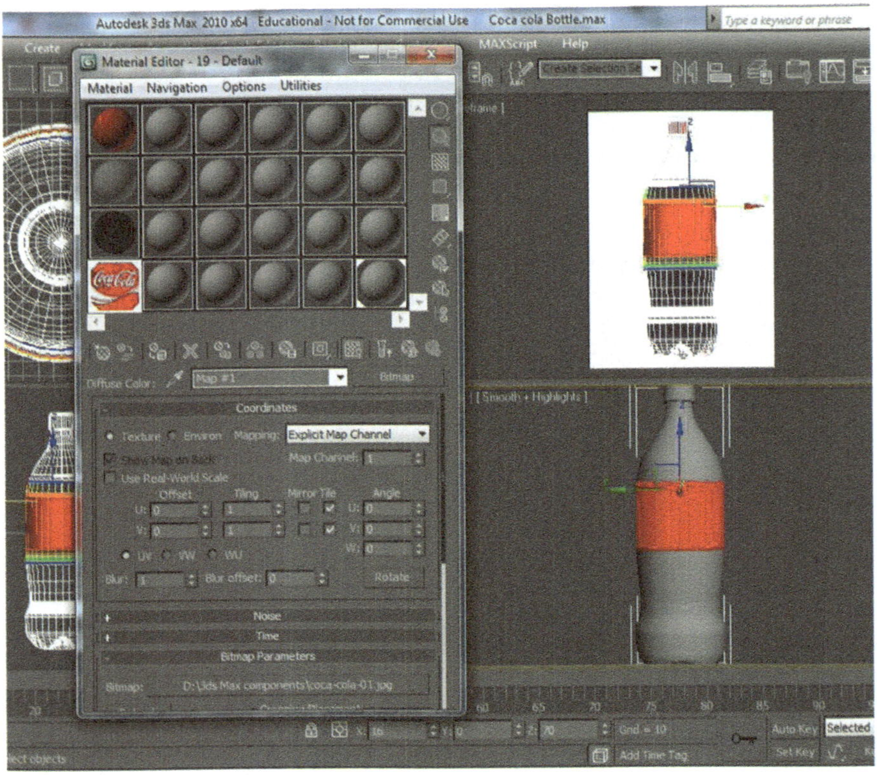

Fig. 11 Material editor

Fig. 12 Complete model with texture mapping

Fig. 13 View of the virtual supermarket

market, while the view of the terrain around the supermarket in Singapore context is shown in Fig. 14.

Fig. 14 View of the terrain

4 Design of Supermarket Serious Game

This virtual supermarket serious game uses Unity3D engine as the game development platform. To start off, game models created in Sect. 3 are imported into Unity3D engine, by dragging corresponding assets files into the Project panel. Unity3D can support imported files in the format of *FBX*, *3ds*, *obj* and *dxf*. Game models imported can be scaled and positioned in the desired orientation under the Inspector panel. The Inspector panel includes all the transform, mesh renderer, script and sound of the game object selected. Scripts can be also attached to desired game objects. With these game models, various game scenes can be created next in Unity3D.

In addition, Unity3D allows animation to be either imported or created in the workspace. In this supermarket serious game, animations are created in Unity3D using timeframe tool.

4.1 Process of Game Design

There are three major parts in developing this serious game: modelling, scene creations and testing. All models are built in Autodesk 3ds MAX stated in Sect. 3. The challenges faced in modelling are creating models as close to reality as possible, resembling physical appearances yet keeping the number of polygons to minimum. It is to ensure the serious game running smoothly.

Game scenes creations and programming are done in Unity3D engine. It enables gamers to interact with the serious game freely. It has taken into consideration the possible thoughts and thinking process of children with ASD. The Kinect interactive functions are combined into Unity3D game scenes, where gamers interact with virtual world using body motions. Microsoft Kinect sensor is used to detect gamers' body movements. Many sets of gestures were tested during the game programming stage. Eventually, a set of simple and easily related gestures is selected for this serious game, shown in Table 1. To bridge Kinect sensor to Unity3D engine, the Flexible

Table 1 Profile conditions of all participants

Output	Gesture
Move down	Left hand below left hip by at least 10 cm
Move down	Right hand below right hip by at least 10 cm
Move up	Left hand above head by at least 15 cm
Move up	Right hand above head by at least 15 cm
Move left	Left hand to the left of left shoulder by at least 10 cm
Move right	Right hand to the right of right shoulder by at least 10 cm
Selection	Left hand apart of right hand by at most 5 cm

Action and Articulation Skeleton Toolkit (FAAST) is used to support gesture controls through Microsoft Kinect in the serious game.

The testing of the serious game is conducted by an experiment with children from the special needs school, which will be presented in Sect. 5.

4.2 Game Implementation

At the start of the serious game, an animation created in Unity3D displays a neighbourhood environment, till bringing players into the entrance of a supermarket. Upon entering the supermarket, three categories of products are displayed for players to select (Fig. 15). Players have to maintain their postures as shown in Fig. 15 during the game, except for executing a command.

Players can use their left hands for left gesture or right hands for right gesture to switch their selection of categories. The default selection is the category at extreme left. The category, which is currently selected is identified by a brighter and larger appearance.

Figure 16 shows the gesture of placing two palms close to each other. It indicates the command of confirming the selection. Upon confirmation, it enters to an animation, which brings gamers to the shelves.

At the shopping shelves, players can select products they would like to purchase. By default, the first product on the left of the top shelve will be pre-selected. From there, players can use their left hands' or right hands' gesture to move left or right to select the next product.

Players can also select products by moving up and down in the shelves. Figure 17 shows the command for selecting the product below the current selection. This

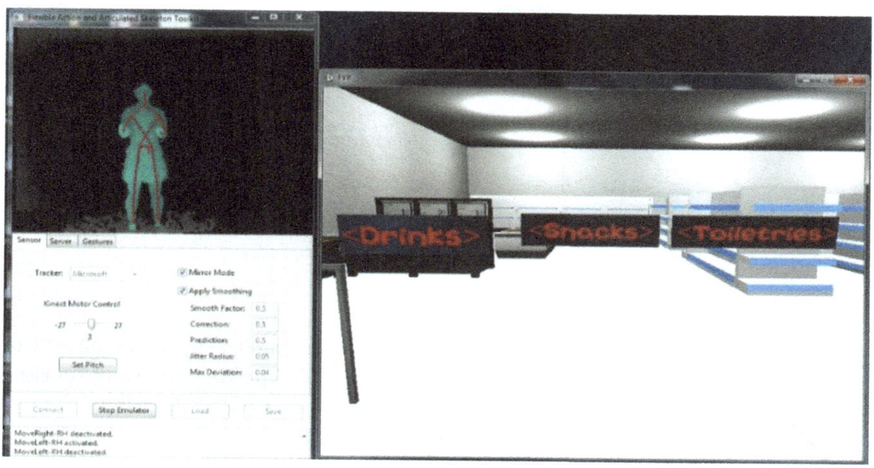

Fig. 15 Main menu with three categories of products

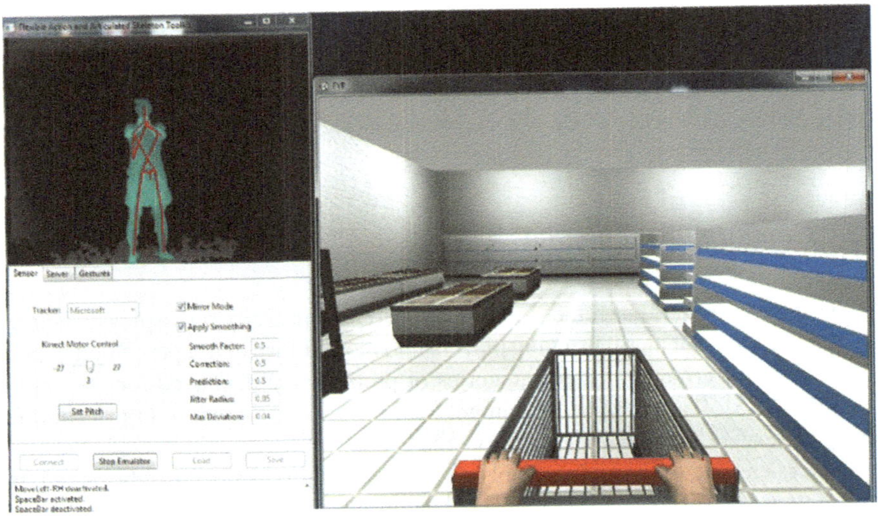

Fig. 16 Animation to bring players to the shelves

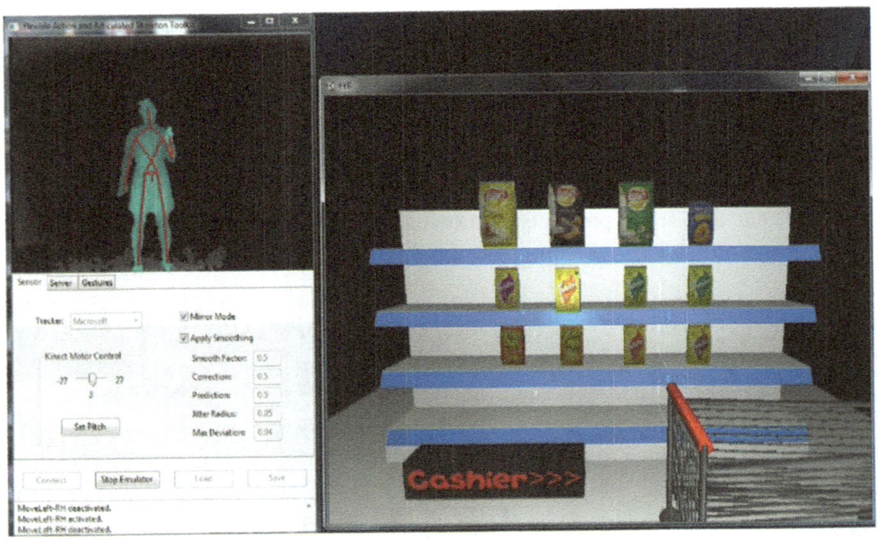

Fig. 17 Selecting product below the current selected item

command can be executed by either the left or right hand. The command turns true when either hand is placed below hip area. To select a product above the current selection, the command is to place either left or right hand above the head.

To confirm the selection, players have to place both palms close together, signifying the grabbing of the product. The selected product will then be positioned nearer to the players. Players can then use right hand gesture to shift the selected product into the trolley.

Players can continue shopping by selecting other intent products and moving them to the trolley one by one. To end the shopping, players need to select the cashier icon at the bottom left by gestures. Then, the confirmation gesture is executed to confirm the selection. It will end the serious game with an animation, which brings players to the cashier counter. At the cashier counter, players need to do the checkout of purchased products, and make the payment with the correct amount.

5 Experiment of Virtual Supermarket Serious Game

5.1 Experiment Objectives

The serious game is to develop independence for children with ASD with the capability of shopping in the supermarket. To find out the effectiveness towards children with moderate ASD, this experiment is conducted in collaboration with AWWA School. The experiment features a virtual environment of the supermarket in a first-person view.

5.2 Experiment Procedures

Two children aged 13 and 14 are chosen for this experiment. Both boys are diagnosed with moderate ASD. The experiment is carried out with the consent of their parents.

The experiment is conducted in one of the classrooms in AWWA School. The serious game is displayed onto a 65-inch television with a Microsoft Kinect sensor located just below the television. The area in front of the television is cleared for the conduction of the experiment. A marker approximately two metres away from the television is placed on the floor to indicate the active zone for the experiment.

5.2.1 Preparation Stage

Both participants attend a briefing session, 30 min before conducting the experiment. During the briefing session, the objectives of this experiment are explained. The gesture commands needed for the experiment have been taught to participants. A demonstration of the virtual supermarket serious game is shown in the briefing session to prepare participants for the experiment.

5.2.2 Experiment Stage

The experiment is conducted twice over a period of two weeks. Each participant goes through two sessions in each experiment. The first session is the practice session for participants to freely try out. The second session is the assessment session. Participants take turns to every session conducted individually in every experiment.

Each session takes around 3–5 min. After every session, a 10-min break is given to participants. In the experiment, participants get to tour around and purchase products in the virtual supermarket serious game. Results are recorded after each assessment session.

5.3 Experiment Results and Discussions

Before the conduct of each experiment, a demonstration is given. During the first demonstration, both participants are very focused on the television screen. Not much attention is paid to the gestures and actions performed. Thus, a practice session is conducted to allow participants to familiarise with the gesture commands needed for the experiment. Each participant goes through the practice session once, each of 5 min. Then, participants start their assessment sessions in the experiment.

The experiment results for each assessment sessions are recorded in Table 2. The results correspond to participants' performance in each experiment. The rating scale is from 1 to 4, where 1 = Seldom (25%), 2 = Sometime (50%), 3 = Often (75%), 4 = Always (100%).

The boxes shaded in bold indicate the improvements by participants by the end of the two experiments.

Figure 18 shows two participants in the experiment of virtual supermarket serious game. Participant 1 displays excitement and interest in the virtual supermarket experiment. He can stay calm and perform each gesture slowly. He manages to identify products on shelves and is able to name what he wants. He performs gestures with some promptings by teachers initially, but is able to execute the rest of the gestures subsequently on his own. He claps after each successful selection that is placed into the trolley, implying the sense of satisfactory and confidence gained during the gameplay.

In the second experiment, Participant 1 shows consistency in his performance. He can name the specific products on shelves, which he wants. He performs gestures independently with minimum prompting. For example, he mentions "Classic" flavour for "Lays Classic"—the product displayed on the top left of the shelf. He stays calm and has displayed a sense of satisfactory throughout the entire experiment. He understands the meaning of each gesture. He waits quietly for his turn and even cheers for the other participant during both experiments.

Participant 2 shows a slightly lower level of concentration as compared to Participant 1. He can identify products after some promptings by teachers. He is able to

Table 2 Experiment results

	1 = Seldom(25%) 2 = Sometime (50%) 3 = Often (75%) 4 = Always (100%)	Participant 1		Participant 2	
No	Objectives	1st experiment	2nd experiment	1st experiment	2nd experiment
1	Attention—able to maintain focus with task for a period of 5 to 8 min	2	**4**	2	**4**
2	Attention—able to maintain pauses between each interaction	2	**3**	1	**3**
3	Sign and Symbol—able to demonstrate / explain meaning of hand movements	2	**3**	1	**3**
4	Effort—able to make an effort to be precise when gathering information (during interactions)	2	**3**	1	**2**
5	Self-Esteem—able to display confident in the experiment	2	**4**	1	**2**
6	Show interest in the virtual supermarket	2	**4**	2	2
7	Tolerance—able to wait for turns and interactions	2	**3**	2	**3**
8	Concept—able to understand the purpose of the experiment	2	**3**	1	**2**
9	Self-esteem—satisfactory level	2	**4**	1	**2**
10	Opinion—How do you feel about the supermarket? 1—Bad 2—Average 3—Good 4—Very Good	3	3	2	**3**

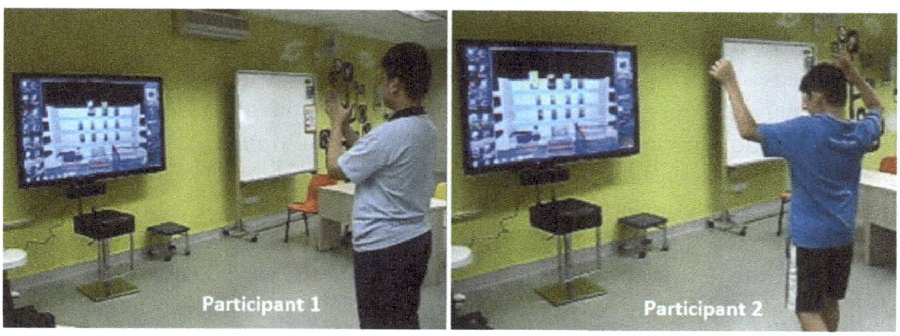

Fig. 18 Participants in the experiment

name products he wants. But he does not display much confidence during the experiment as he quickly rushes back to his seat after each session. This could reflect the low self-esteem in Table 2. He would probably require more time to adapt to a new environment. He performs gestures with the guidance of teachers. He can maintain pauses between each gesture. He waits patiently for his turn after each session.

In the second experiment, Participant 2 displays calmness and confidence. He can execute gestures in the experiment with less guidance from teachers. He shows a clear understanding of the meaning of each gesture.

From the experimental results, Participant 1 seems to adapt to the virtual supermarket much faster. He displays confidence and satisfaction during both experiments. Though Participant 2 displays a slightly lower confidence level during his first experiment, large improvement is observed in his second experiment. He is also able to perform the experiment with little prompting, in addition, improvement is shown in this participant over a short period of time. This result shows that children with moderate ASD are able to benefit from this serious game. Although it may not necessarily prove that the skills in the virtual supermarket will be adopted fully in reality, it can surely serve as an aid to teach intervention and social skills to children with ASD.

Due to the time constraints in the tight schedule of AWWA School, the current experiment duration is short. Moving forward, the conduct of the experiment should be performed over a longer period to justify the accurate effect of the virtual supermarket serious game. It will ensure the skills needed can be picked up by children with ASD through the experience and gameplay.

6 Conclusions

Children with ASD may not be independent enough for shopping alone in crowded supermarkets. A virtual supermarket serious game is developed in this work to train children with ASD on life skills—shopping in supermarket. The virtual supermarket serious game can better prepare the children with ASD for the real supermarket experience in the future. Experiment has been conducted in AWWA School, a special needs school in Singapore. The experiment results show the performance improvements of participants over two experiment sessions across 2 weeks.

Although the experiment results may not be enough to fully justify the benefits of this serious game towards developing a new skill for children with ASD, it definitely has the ability to serve as a visual aid to familiarise them with supermarket environments. It indicates that children with ASD are able to adapt the use of the body gestures to deliver the interactive capabilities in the virtual environment. As a result, this serious game is viewed feasible for the development of independence in children with ASD.

Acknowledgements The authors would like to thank the students, teachers, staffs, principal and parents of AWWA School for their support, help and feedback in this research work.

References

1. Brentani, H., Paula, C., Bordini, D., Rolim, D., Sato, F., Portolese, J., Pacifico, M., McCracken, J.: Autism spectrum disorders: an overview on diagnosis and treatment. Braz. J. Psychiat. **35**(1), 62–72 (2013). https://doi.org/10.1590/1516-4446-2013-S104
2. Bourazeri, A., Bellamy-Wood, T., Arnab, S.:EnCity: a serious game for empowering young people with Down's syndrome. In: 5th IEEE International Conference on Serious Games and Applications for Health (2017). https://doi.org/10.1109/SeGAH.2017.7939267.
3. Gialloreti, L.E., Curatolo, P.: Autism spectrum disorder: why do we know so little? Front. Neurol. **9**, 670 (2018). https://doi.org/10.3389/fneur.2018.00670
4. Johnston, D., Egermann, H., Kearney, G.: SoundFields: a virtual reality game designed to address auditory hypersensitivity in individuals with autism spectrum disorder. Appl. Sci. **10**, 2996 (2020). https://doi.org/10.3390/app10092996
5. Keeling, K., Myles, B.S., Gagnon, E., Simpson, R.L.: Using the power card strategy to teach sportsmanship skills to a child with autism. Focus Autism Dev. Disabil. **18**(2), 105–111 (2003)
6. Mesa-Gresa, P., Gil-Gómez, H., Lozano-Quilis, J.A., Gil-Gómez, J.A.: Effectiveness of virtual reality for children and adolescents with autism spectrum disorder: an evidence-based systematic review. Sensors (Basel, Switzerland) **18**(8), 2486 (2018). https://doi.org/10.3390/s18082486
7. Parsons, S., Mitchell, P.: The potential of virtual reality in social skills training for people with autistic spectrum disorders. Intell. Disabil. Res. **46**(Pt 5), 430–443 (2002). https://doi.org/10.1046/j.1365-2788.2002.00425.x
8. Schreiber, C.: Social skill interventions for children with high-functioning autism spectrum disorders. J. Positive Behavior Intervent. **13**(1), 49–62 (2011). https://doi.org/10.1177/1098300709359027
9. Suparjoh, S., Shahbodin, F., Mohd, C.K.N.: Technology-assisted intervention for children with autism spectrum disorder using augmented reality. Int. J. Recent Technol. Eng. **8**(5) (2020)
10. Valencia, K., Rusu, C., Quiñones, D., Jamet, E.: The impact of technology on people with autism spectrum disorder: a systematic literature review. Sensors (Basel, Switzerland) **19**(20), 4485 (2019). https://doi.org/10.3390/s19204485
11. Yuan, S., Ip, H.: Using virtual reality to train emotional and social skills in children with autism spectrum disorder. Lond. J. Primary Care **10**(4), 110–112 (2018). https://doi.org/10.1080/17571472.2018.1483000

Afterword

E-inclusion: Equity and Equality by Electronic Learning

The 'Salamanca Statement and Framework for Action on Special Needs Education' appeared 25 years ago in the field of special education [1]. Arguably be seen as the most significant international document that has ever been published in this domain, this publication has pushed inclusive education throughout Europe.

Providing the diverse educational needs of students is one of the most challenging educational goals in today's schools. In Europe, inclusive education is supported by European Commission funding and promoted jointly by the European Agency for Special Needs and Inclusive Education and by the United Nations Educational, Scientific and Cultural Organization. Many member states of the European Union have the implementation of inclusive education high on their educational agenda. The core underlying assumption of this policy for children with Special Educational Needs (SEN) is that they would benefit most from education alongside neuro-typical children in mainstream schools, as opposed to special schools catering specifically for those with SEN. However, inclusive educational practice is often restricted to mere placement of students with SEN in mainstream schools and to remedial teaching practices. In the first case, students that would normally attend special schools are enrolled in mainstream education. In the second case, mainstream education provides support trajectories for these SEN students [2]. In both cases, students are defined by their deficiencies [2, 3].

Research on SEN has been undertaken from two distinct perspectives that resonate with two perspectives on inclusion. The positivist medical perspective, as described above, on one hand, where students with SEN are perceived to be deficient, relative to their typically developing peers. Specialized individual support, like counseling or remedial teaching, is aimed at treatment of students with SEN. The interpretivist equality perspective, on the other hand, states that it is mostly the environment that produces inequalities and needs [4, 5]. Support is about changing the perceptions and attitudes of actors within the educational environment, in order for students

Y. Cai and Q. Cao (eds.), *When VR Serious Games Meet Special Needs Education*, Gaming Media and Social Effects, https://doi.org/10.1007/978-981-33-6942-9

to reach their full potential [2, 3, 6]. These changing conceptions regarding the notion of inclusive education have outgrown the fierce discussions on themes like specialist segregated settings, integration, and mainstreaming [7]. In a more profound understanding of inclusive practice, schools aim to develop a supporting environment for all students, and an inclusive school community that nourishes the change needed to become more inclusive [2].

Within this inclusive learning environment, assistive technology and emerging technologies have been widely identified as an important support and facilitator for students with SEN and, in general, persons with disabilities to enhance greater participation in educational settings and in their daily lives. I am coining this movement 'E-Inclusion' in order to support the mental health of all students with SEN through an emphatic lens of equality and equity by electronic learning. All students with SEN should have instruction in the use of appropriate assistive technology from an educator or teacher. Instruction should be goal-oriented; focused on academic, vocational, and independent living skills; and should build on a hierarchy of skills. However, teachers do not necessarily integrate technology into their classroom teaching practices. This is where this book can fill this void.

This book features the efforts of Dr. Yiyu Cai and his team at Nanyang Technological University in developing Virtual Reality (VR), Augmented Reality (AR), Simulation and Serious Games for Special Needs Education and students with SEN in Singapore. Around 2008, Dr. Cai started developing VR Serious Games for Special Needs Education. Over the past 10 years, together with his students, they have designed and developed a good number of projects for children with autism or other disabilities by working closely with local special needs schools such as AWWA and METTA. The book consists of ten chapters and an introductory chapter with focuses on game-assisted learning of living skills, communication skills, and job skills. This book shares the principles and methodologies in detail to design and develop VR serious games for special needs education. Some of the games detailed in the book include VR Showering, Road Crossing, School Bag Packing, and so on. Several tools are also introduced for project development. In addition to technical development of VR serious games, this book has several chapters describing evaluations of the games through experiment design and analysis.

I first met Dr. Cai in 2010, when I was a visiting professor with National Institute of Education (NIE), at Academic Group of Early Childhood and Special Needs Education. Since then, we collaborate in this cutting-edge field of (assistive) technology, students with SEN, and special education. We have worked together on a number of projects and activities including: (1) development of the virtual pink dolphins game for children with autistic spectrum disorder; (2) joint establishment of the Europe-Asia Symposium on Simulation and Serious Games in which we co-chaired four times the symposia in Singapore, the Netherlands, China, and Finland. We are going to organize the 5th version of the conference online end of this year, 2020; (3) co-editing of two books on Simulation and Serious Games published in Springer, and (4) collaboration on promoting Europe-Asia partnership particularly in Special Needs Education by helping schools in Asia and Europe to establish international collaboration, visits, and exchanges. I am cherishing my companionship with Dr.

Cai, both in a personal and professional way, and hope we can continue collaborating in this important field. I encourage readers to take notice of the developments described in this book and as such I find this book highly relevant to the field of inclusive and special education regarding the use of assistive technology and emerging technologies.

October 2020

Sui Lin Goei (Ph.D.)
Professor, VU Amsterdam and
Windesheim University of Applied Sciences
The Netherlands

References

1. Ainscow, M., Slee, R., Best, M.: Editorial: the Salamanca Statement: 25 years on. Int. J. Incl. Educ. **23**(7–8), 671–676 (2019). https://doi.org/10.1080/13603116.2019.1622800
2. Nilholm, C.: Research about inclusive education in 2020 – how can we improve our theories in order to change practice? Eur. J. Spec. Needs. 1–13 (2020). https://doi.org/10.1080/08856257.2020.1754547
3. Terzi, L.: Reframing inclusive education: Educational equality as capability equality. Camb. J. Educ. **44**(4): 479–493 (2014). https://doi.org/10.1080/0305764X.2014.960911
4. Ainscow, M.: Promoting inclusion and equity in education: lessons from international experiences. Nord. J. Stud. Educ. Policy. **6**(1), 7–16 (2020). https://doi.org/10.1080/20020317.2020.1729587
5. Qu, X.: A critical realist model of inclusive education for children with special educational needs and/or disabilities. Int. J. Incl. 1–15 (2020). https://doi.org/10.1080/13603116.2020.1760366
6. Avramidis, E., Norwich, B.: Teachers' attitudes towards integration/inclusion: a review of the literature. Eur. J. Spec. Needs Educ. **17**(2), 129–147 (2002). https://doi.org/10.1080/08856250210129056
7. Meijer, C., Watkins, A.: Changing conceptions of inclusion underpinning education policy. In book: Implementing Inclusive Education: Issues in Bridging the Policy-Practice Gap, pp.1–16 (2016). https://doi.org/10.1108/S1479-363620160000008001

Index

Y. Cai and Q. Cao (eds.), *When VR Serious Games Meet Special Needs Education*, Gaming Media and Social Effects, https://doi.org/10.1007/978-981-33-6942-9

D

E

The manufacturer's authorised representative in the EU is Springer Nature Customer Service Centre GmbH, Europaplatz 3, 69115 Heidelberg, Germany. If you have any concerns regarding our products, please contact ProductSafety@springernature.com

Printed and bound by CPI Group (UK) Ltd, Croydon, CR0 4YY
30/09/2026
02225320-0004